HOLT BIOSOURCES LAB PROGRAM

# BIOTECHNOLOGY

## Teacher's Edition

INCLUDES LABS D1–D12

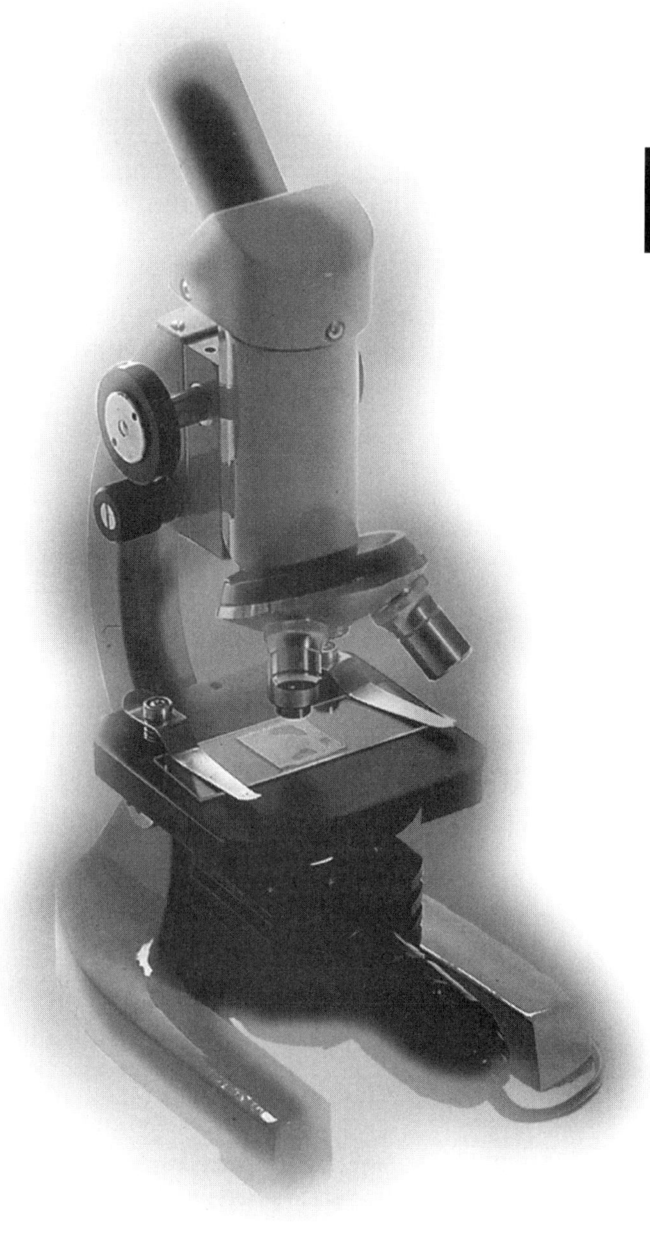

HOLT, RINEHART AND WINSTON
*Harcourt Brace & Company*
**Austin** • New York • Orlando • Atlanta • San Francisco • Boston • Dallas • Toronto • London

HOLT BIOSOURCES *LAB PROGRAM*

# BIOTECHNOLOGY


## Staff Credits

**Editorial Development**

Carolyn Biegert
Janis Gadsden
Debbie Hix

**Copyediting**

Amy Daniewicz
Denise Haney
Steve Oelenberger

**Prepress**

Rose Degollado

**Manufacturing**

Mike Roche

**Design Development and Page Production**

Morgan-Cain & Associates

## Acknowledgments

**Contributors**

David Jaeger
Will C. Wood High School
Vacaville, CA

George Nassis
Kenneth G. Rainis
WARD'S Natural Science Establishment
Rochester, NY

Suzanne Weisker
Science Teacher and Department Chair
Will C. Wood High School
Vacaville, CA

**Editorial Development**

WordWise, Inc.

**Cover**

Design—Morgan-Cain & Associates
Photography—Sam Dudgeon

**Lab Reviewers**

***Lab Activities***
Ted Parker
Forest Grove, OR

Mark Stallings, Ph.D.
Chair, Science Department
Gilmer High School
Ellijay, GA

George Nassis
Kenneth G. Rainis
Geoffrey Smith
WARD'S Natural Science Establishment
Rochester, NY

***Lab Safety***
Kenneth G. Rainis
WARD'S Natural Science Establishment
Rochester, NY

Jay Young, Ph.D
Chemical Safety Consultant
Silver Spring, MD





Printed in the United States of America

**ISBN 0-03-051408-8**

1 2 3 4 5 6 022 00 99 98 97

**BIOTECHNOLOGY**

HOLT BIOSOURCES LAB PROGRAM

# *Contents*

***Teacher's Edition***

# Contents, *continued*

## Unit 5 *Viruses, Bacteria, Protists, and Fungi*

# Maintaining a Safe Lab

## Building Safety Partnerships: You're Not Alone

A safe laboratory can only be achieved through a partnership among all parties concerned, not just among students or teachers. Materials for teachers and students need to be thorough, explicit, and persuasive. Teachers must actively boost safety consciousness among students, fellow faculty members, administrators, and parents. For success, everyone must agree to respect the same laboratory rules, to obtain and use the proper safety equipment, and to take appropriate precautions during a lab activity.

An excellent way to start building this safety partnership with your students is to use the Safety Contract on the **Holt BioSources Teaching Resources CD-ROM.** Have each student fill out a contract and return it to you. Keep the contracts on file, in case you need to remind students of their promises.

## Where to Start

In each lab activity, safety symbols are included and specific safety procedures are highlighted where appropriate. Detailed descriptions of each safety symbol and hazards and precautions related to each one can be found in your biology textbook and in the "Laboratory Safety" section of the pupil's edition of the *Inquiry Skills Development, Laboratory Techniques and Experimental Design,* and *Biotechnology* manuals. The safety symbol descriptions are included in expanded form in the following section.

The expanded safety symbol descriptions and other information included in the following sections will help you plan and maintain a safe and healthy laboratory environment.

**This information is not all-inclusive. Each school's lab situation is different, and no publication could list safe practices for all situations that could possibly arise.**

Be sure that you are aware of any federal, state, or local laws that may cover your lab. Although laws and regulations can vary from place to place and from time to time, you can build a safe program suited to your situation using this information.

# Safety Symbols

## Eye Safety

- **Wear approved chemical safety goggles as directed.** Goggles should always be worn whenever you or your students are working with a chemical or solution, heating substances, using any mechanical device, or observing physical processes. See the "Safety Equipment" section for specific tips on the types of goggles to be worn.

  Teachers should model appropriate behavior by wearing safety goggles when appropriate. Some teachers have success when they turn this into a game: students found without safety goggles must wear them during the next lecture period. If students catch the teacher not wearing safety goggles, the teacher must wear them during the next lecture period.

- **In case of eye contact, go to an eyewash station and flush eyes (including under the eyelids) with running water for at least 15 minutes.** The teacher or other adult in charge must be notified immediately.
- **Wearing contact lenses for cosmetic reasons is prohibited in the laboratory.** Be sure to make this clear to students at the beginning of the year. First, take a poll of contact-lens wearers. Explain the precautions necessary, noting that liquids or gases can be drawn up under a contact lens and onto the eyeball. If a student must wear contact lenses prescribed by a physician, be sure the students wear approved eyecup safety goggles, which are similar to goggles worn for swimming.
- **Never look directly at the sun through any optical device or lens system, and never gather direct sunlight to illuminate a microscope.** Such actions will concentrate light rays that can severely burn the retina, possibly causing blindness. At the beginning of the year, make sure each microscope you will use has an appropriate and functioning light source.

## Electrical Supply

- **Be sure you know the location of the master shut-off for all circuits and other utilities in the lab.** If the circuit breakers are locked up, make sure you have a key in case of an emergency. Color code or label the necessary switches. Be sure you will remember what to do under the pressures of an emergency.
- **Be sure all outlets have correct polarity and ground-fault interruption.** Polarity can be tested with an inexpensive (about $5.00) continuity tester available from most electronic hobby shops. Use only electrical equipment with three-prong plugs and three-wire cords. Each electrical socket in the laboratory must be a three-holed socket with a GFI (ground fault interrupter) circuit. In some cases, rewiring the lab may be necessary. Be sure your supervisors understand the potential hazards and costs of leaving the lab in an unsafe configuration.
- **Electrical equipment should be in the "off" position before it is plugged into a socket.** After a lab activity is completed, the equipment should first be turned off and then unplugged. Wiring hookups should not be made or altered except

when an apparatus is disconnected from its AC or DC power source and the power switch, if applicable, is off.

- **Do not let electrical cords dangle from work stations.** Dangling cords are a hazard that can cause tripping or electrical shock.
- **Tape electrical cords to work surfaces.** This will prevent falls and decrease the chances of equipment being pulled off the table. If you find you have too many cords to be taped down, that could be a sign of a poorly designed lab, which will be prone to other problems as well. Sometimes simply rearranging the lab desks alleviates some flaws.
- **Never use or allow students to use equipment with frayed or kinked cords.** Check all of the electrical equipment at the beginning and end of each year. It is better to omit an activity because the equipment is unsafe than to proceed with an activity that results in an injury to yourself or a student.
- **Never use electrical equipment around water or with wet hands or clothing.** The area under and around electrical equipment should be dry. Electrical cords should not lie in puddles of spilled liquid.
- **Use dry cells or rechargeable batteries as direct current (DC) sources.** Do not use automobile storage batteries or AC-to-DC converters; these two sources of DC current can present serious electrical shock hazards. When storing dry cells and rechargeable batteries, cover both terminals with insulating tape.
- **Before leaving the laboratory, be sure all electrical equipment has been turned off and is unplugged.**

## *Clothing Protection*

- **Wear a lab apron or lab coat when working in the laboratory to prevent chemicals or chemical solutions from contacting skin or contaminating clothes.** Suggested styles of lab aprons are discussed in the "Safety Equipment" section that follows. Be sure students confine all loose clothing and long jewelry. Open-toed shoes should not be allowed in the laboratory.

## *Animal Care*

- **Do not touch or approach any animal in the wild.** Be sure you and your students are aware of any poisonous or dangerous animals in any area where you will be doing fieldwork.
- **Always insist that students obtain your permission before bringing any animal (or pet) into the school building.** There are legitimate reasons to bring animals to school, but be certain that such an occasion does not present a danger or a distraction to students.
- **Handle all animals with proper caution and respect.** Mishandling or abuse of any animal should not be tolerated. The National Association of Biology Teachers guidelines for the use of live animals, reproduced in the "Animal Care" section, provide a good framework for planning specific procedures.

## *Sharp Object Safety*

- **Use extreme care with all sharp instruments, such as scalpels, sharp probes, and knives.** You may want to consider restricting the use of such objects to lab activities for which there are no substitutes.
- **Never use double-edged razors in the laboratory.**
- **Never cut objects while holding them in your hand.** Place objects on a suitable work surface. Be sure your lab has an adequate supply of dissecting pans and similar surfaces for cutting.

## *Chemical Safety*

More detailed information on chemical hazards, the use of MSDSs (Material Safety Data Sheets), and safe chemical storage can be found in the "Reagents and Storage" and "Chemical Handling and Disposal" sections that follow.

- **Always wear appropriate personal protective equipment.** Safety goggles, gloves, and a lab apron or lab coat should always be worn when working with any chemical or chemical solution.
- **Never taste, touch, or smell any substance or bring it close to your eyes, unless specifically directed to do so.** If students need to note the odor of a substance, have them do so by waving the fumes toward themselves with their hands. Make sure there are enough suction bulbs for any pipetting that needs to be done. Set a good example, and never pipet any substance by mouth.
- **Always handle any chemical or chemical solution with care.** Nothing in the lab should be considered harmless. Even nontoxic substances can easily be contaminated. Check the MSDS for each chemical prior to a lab activity, and observe safe-use procedures. Be sure to have appropriate containers available for unused reagents so that students won't return them to reagent bottles. Store chemicals according to the directions in the "Reagents and Storage" section.
- **Never mix any chemicals unless you are certain about what you are doing and why.** Many common chemicals react violently with each other. Consult section V of a chemical's MSDS for compatibility information.
- **Never pour water into a strong acid or base.** The mixture can produce heat and splatter. Remember this rhyme:

  "Do as you oughta—  
  Add acid (or base) to water."

- **Have a spill-control plan and kit ready.** Students should not handle chemical spills. Be sure that your spill-control kit contains neutralizing agents, sand, and other absorbent material.
- **Check for the presence of any source of flames, sparks, or heat (open flame, electrical heating coils, etc.) before working with flammable liquids or gases.**

## *Plant Safety*

- **Do not ingest any plant part used in the laboratory (especially seeds sold commercially).** Commercially sold seeds often are coated with fungicidal agents. Do not rub any sap or plant juice on your eyes, skin, or mucous membranes.
- **Wear protective (disposable polyethylene) gloves when handling any wild plant.**
- **Wash hands thoroughly after handling any plant or plant part (particularly seeds).** Avoid touching your hands to your face and eyes.
- **Do not inhale or expose yourself to the smoke of any burning plant.** Some irritants travel in smoke and can cause inflammation in the throat and lungs.
- **Do not pick wildflowers or other plants unless permission from appropriate authorities has been obtained in advance.**

## *Proper Waste Disposal*

- **Have students clean and decontaminate all work surfaces and personal protective equipment after each lab activity.** Prompt and frequent cleaning helps keep contamination problems to a minimum.
- **Set aside special containers for the disposal of all sharp objects (broken glass and other contaminated sharp objects) and other contaminated materials (biological or chemical).** Make sure these items are disposed of in an environmentally sound way.

## *Hygienic Care*

- **Keep your hands away from your face and mouth.**
- **Wash your hands thoroughly before leaving the laboratory.** Have bactericidal soap available for students to use.
- **Remove contaminated clothing immediately; launder contaminated clothing separately.** Have a few spare T-shirts and shorts or sweatsuits available in case of an emergency involving clothing.
- **Demonstrate the proper techniques when handling bacteria or microorganisms.** Examine microorganism cultures (such as those in petri dishes) without opening them.
- **Collect all stock and experimental cultures for proper disposal.** See the "Safety With Microbes" and "Release and Disposal of Organisms" sections for instructions on materials and cultures used in these lab activities.

## *Heating Safety*

- **When heating chemicals or reagents in a test tube, never point the test tube toward anyone.**
- **Use hot plates, not open flames.** Be sure hot plates have an On-Off switch and indicator light. Never leave hot plates unattended, even for a minute.

  Check all hot plates for malfunctions several times during the school year. Never use alcohol lamps.
- **Know the location of laboratory fire extinguishers and fire blankets.** Have ice readily available in case of burns or scalds. Make certain that your laboratory fire extinguishers are tri-class (A-B-C) and are useful for all types of fires.
- **Use tongs or appropriate insulated holders when heating objects.** Heated objects often do not look hot. Set a good example by using tongs or other holders to handle an object whenever there is a possibility that the object could be warm.
- **Keep combustibles away from heat and other ignition sources.**

## *Hand Safety*

- **Never cut objects while holding them in your hand.**
- **Wear protective gloves when working with stains, chemicals, chemical solutions, or wild (unknown) plants.**

## *Glassware Safety*

- **Inspect glassware before use; never use chipped or cracked glassware.** Use borosilicate glass for heating. Check all glassware several times a year, and discard anything that shows signs of chipping or cracking.
- **Hold glassware firmly, but do not squeeze it.** Glass is fragile and may break if it is not handled carefully. Be sure your hands and the glassware are dry when you are handling glassware.
- **Do not attempt to insert glass tubing into a rubber stopper without taking proper precautions.** Lubricate the stopper and the glass tubing. Use heavy leather gloves to protect your hands from shattering glass. To prevent puncture wounds, be sure your hand is clear of the hole where the glass tubing will emerge.
- **Always clean up broken glass by using tongs and a brush and dustpan.** Discard the pieces in an appropriately labeled "sharps" container.

## *Safety With Gases*

- **Never directly inhale any gas or vapor.** Do not put your nose close to any substance having an odor.
- **Be sure that your lab has excellent ventilation.** Some work will still require a chemical fume hood. If your lab does not have good ventilation, investigate opportunities to improve it. Be certain your supervisors are aware of potential hazards due to ineffective ventilation.

# Laboratory Rules

Post the following rules in the laboratory, and discuss them with students. Afterward, give students the "Safety Quiz" on the **Holt BioSources Teaching Resources CD-ROM.**

- **Never work alone in the laboratory.**
- **Never perform any experiment not specifically assigned by your teacher.** Never work with any unauthorized material.
- **Never eat, drink, or apply cosmetics in the laboratory.** Never store food in the laboratory. Keep hands away from faces. Wash your hands at the conclusion of each laboratory investigation and before leaving the laboratory. Remember that some hair products are highly flammable, even after application.
- ***NEVER* taste chemicals. *NEVER* touch chemicals.** Even common substances should be considered dangerous, since they can be easily contaminated in the lab.
- **Do not wear contact lenses in the lab.** Chemical vapors can get between the lenses and the eyes and cause permanent eye damage.
- **Know the location of all safety and emergency equipment used in the laboratory.** Examples include eyewash stations, safety blankets, safety shower, fire extinguisher, first-aid kit, and chemical-spill kit.
- **Know fire drill procedures and the locations of exits.**
- **Know the location of the closest telephone,** and be sure there is a posted list of emergency phone numbers, including the poison control center, fire department, police department, and ambulance service.
- **Familiarize yourself with a lab activity—especially safety issues—before entering the lab.** Know the potential hazards of the materials and equipment to be used and the procedures required for the activity. Before you start, ask the teacher to explain any parts you do not understand.
- **Before beginning work: tie back long hair, roll up loose sleeves, and put on any personal protective equipment as required by your teacher.** Avoid wearing loose clothing or confine loose clothing that could knock things over, ignite from flame, or soak up chemical solutions. Do not wear open-toed shoes to the lab. If there is a spill, your feet could be injured.
- **Report any accidents, incidents, or hazards—no matter how trivial—to your teacher immediately.** Any incident involving bleeding, burns, fainting, chemical exposure, or ingestion should also be reported to the school nurse or physician.
- **In case of fire, alert the teacher and leave the laboratory.**
- **Keep your work area neat and uncluttered.** Bring only the books and materials needed to conduct a lab activity. Stay at your work area as much as possible. The less movement in a lab, the fewer spills and other accidents that can occur.
- **Clean your work area at the conclusion of a lab activity as your teacher directs.**
- **Wash your hands with soap and water after each lab activity.**

# Safety Equipment

## Do You Have What It Takes?

- **Chemical goggles** (meeting ANSI [American National Standards Institute] standard Z87.1): These should be worn when working with any chemical or chemical solution other than water, when heating substances, when using any mechanical device, or when observing physical processes that could eject an object.

  Wearing contact lenses for cosmetic reasons should be prohibited in the laboratory. If a student must wear contact lenses prescribed by a physician, that student should wear eyecup safety goggles meeting ANSI standard Z87.1 (similar to swimmers' goggles).

- **Face shield** (meeting ANSI standard Z87.1): Use in combination with safety goggles when working with corrosives.

- **Eyewash station:** The station must be capable of delivering a copious, gentle flow of water to both eyes for at least 15 minutes. **Portable liquid supply devices are not satisfactory and should not be used.** A plumbed-in fixture or a perforated spray head on the end of a hose attached to a plumbed-in outlet and designed for use as an eyewash fountain is suitable if it meets ANSI standard Z358.1 and is within a 30-second walking distance from any spot in the room.

- **Safety shower** (meeting ANSI standard Z358.1): Location should be within a 30-second walking distance from any spot in the room. Students should be instructed in the use of the safety shower for a fire or chemical splash on their body that cannot be simply washed off.

- **Gloves:** Polyethylene, neoprene, or disposable plastic may be used. Nitrile or butyl rubber gloves are recommended when handling corrosives.

- **Apron:** Gray or black rubber-coated cloth or a nylon-coated vinyl halter is recommended.

# Prudent Precautions

What would you do if a student dropped a liter bottle of concentrated sulfuric acid? RIGHT NOW? Are you prepared? Could you have altered your handling and storage methods to prevent or lessen the severity of this incident? PLAN now how to effectively react BEFORE you need to. Planning tips include the following:

1. Post the phone numbers of your regional poison control center, fire department, police department, ambulance service, and hospital ON your telephone.
2. Practice fire and evacuation drills during labs and at all times during the year, not just in the fall. Post an evacuation diagram and an established evacuation procedure by every entrance to the laboratory.
3. Have drills on what students MUST do if they are on fire or experience chemical contact or exposure.
4. Mark the locations of eyewash stations, the safety shower, fire extinguishers (A-B-C tri-class), the chemical-spill kit, the first-aid kit, and fire blankets in the laboratory and storeroom. Make sure you have all of the necessary safety equipment prior to conducting each lab activity, and be certain the equipment is in good working order.
5. Lock your laboratory (and storeroom) when you are not present.
6. Compile an MSDS file for all chemicals. This reference resource should be readily accessible in case of spills or other incidents. (Information about MSDSs is found in the "Reagents and Storage" section.)
7. Develop spill-control procedures. Handle only incidents that you FEEL COMFORTABLE handling. Situations of greater severity should be handled by trained hazardous-material responders.
8. Under no circumstances should students fight fires or handle chemical spills.
9. Be sure to recognize and heed the signal words used on most safety labels for materials, equipment, and procedures:
   CAUTION—low level of risk associated with use or misuse
   WARNING—moderate level of risk associated with use or misuse
   DANGER—high level of risk associated with use or misuse
10. Be trained in first aid and basic life support (CPR) procedures. Have first-aid kits and spill kits readily available.
11. Before the class begins a lab activity, review specific safety rules and demonstrate proper procedures.
12. Never permit students to work in your laboratory without your supervision. No unauthorized investigations should ever be conducted, nor should unauthorized materials be brought into the laboratory.
13. Fully document ANY INCIDENT that occurs. Documentation will provide the best defense in terms of liability, and it is a critical tool in helping to identify area(s) of laboratory safety that need improvement. Remind students that any safety incident, no matter how trivial, must be reported directly to you.

# Safety With Microbes

## What You Can't See CAN Hurt You

Pathogenic (disease-causing) microorganisms are not appropriate investigation tools in the high school laboratory and should never be used.

Consult with the school nurse to screen students whose immune system may be compromised by illness or who may be receiving immunosuppressive drug therapy. Such individuals are extraordinarily sensitive to potential infection from generally harmless microorganisms and should not participate in laboratory activities unless permitted to do so by a physician. Do not allow students with any open cuts, abrasions, or sores to work with microorganisms.

## Aseptic Technique

Demonstrate correct aseptic technique to students PRIOR to conducting a lab activity. Never pipet liquid media by mouth. Wherever possible, use sterile cotton applicator sticks in place of inoculating loops and Bunsen burner flames for culture inoculation. Remember to use appropriate precautions when disposing of cotton applicator sticks: they should be autoclaved or sterilized before disposal.

Treat ALL microbes as pathogenic. Seal with tape all petri dishes containing bacterial cultures. Do not use blood agar plates, and never attempt to cultivate microbes from a human or animal source.

Never dispose of microbe cultures without first sterilizing them. Autoclave or steam-sterilize all used cultures and any materials that have come in contact with them at 120°C and 15 psi for 15–20 minutes. If an autoclave or steam sterilizer is not available, flood or immerse these articles with full-strength household bleach for 30 minutes, and then discard. Use the autoclave or steam sterilizer yourself; do not allow students to use these devices.

Wash all lab surfaces with a disinfectant solution before and after handling bacterial cultures.

## Handling Bacteriological Spills

Never allow students to clean up bacteriological spills. Keep on hand a spill kit containing 500 mL of full-strength household bleach, biohazard bags (autoclavable), forceps, and paper towels.

In the event of a bacterial spill, cover the area with a layer of paper towels. Wet the paper towels with the bleach, and allow to stand for 15–20 minutes. Wearing gloves and using forceps, place the residue in the biohazard bag. If broken glass is present, use a brush and dustpan to collect the broken material and place it in a suitably marked container.

# Reagents and Storage

## General Guidelines

- Store bulk quantities of chemicals in a safe and secure storeroom, not in the teaching laboratory. Store them in well-ventilated, dry areas protected from sunlight and localized heat. Store by similar hazard characteristics, not alphabetically. (See "Chemical Hazard Classes" and "Chemical Storage" for additional recommendations.)
- Label student reagent containers with the substance's name and hazard class(es). Be sure to use labeling materials that won't be affected by the reagent or other chemicals that will be stored nearby. (See "Chemical Hazard Classes" for additional recommendations.)
- Dispose of hazardous waste chemicals according to federal, state, and local regulations. Refer to the Material Safety Data Sheets (available through your supplier) for recommended disposal procedures. Some disposal information is also included in the "Chemical Handling and Disposal" section on pages T21–T26. **NEVER ASSUME** that a reagent can be safely poured down the drain.
- Have a chemical-spill kit immediately available. Know the procedures for handling a spill of any chemical used during a lab activity or in preparing reagents. Never allow students to clean up hazardous chemical spills.
- Remove all sources of flames, sparks, and heat from the laboratory when any flammable material is being used.

## Chemical Record-Keeping and MSDSs

Maintaining a current inventory of chemicals can help you keep track of purchase dates and amounts. In addition, if you use your inventory to determine what you need for the school year, you can cut down on storage problems.

The purpose of a Material Safety Data Sheet (MSDS) is to provide readily accessible information on chemical substances commonly used in the science laboratory or in industry. MSDSs are available from suppliers of chemicals.

MSDSs should be kept on file and referred to BEFORE handling ANY chemical. The MSDSs can also be used to instruct students on chemical hazards, to evaluate spill and disposal procedures, and to warn of incompatibilities with other chemicals or mixtures.

Each MSDS is divided into the following sections:

**I.** Material Identification: includes name, common synonyms, reference codes, and precautionary labeling

**II.** Ingredients and Hazards: identifies dangerous components of mixtures

**III.** Physical Data: includes information such as melting point, boiling point, appearance, odor, density, etc.

IV. Fire and Explosion Hazard Data: includes flash point, description of fire-extinguishing media and procedures, and information on unusual fire and explosion hazards

V. Health Hazard Data: describes problems associated with inhalation, skin contact, eye contact, skin absorption, and ingestion, along with first-aid procedures

VI. Reactivity Data: includes information on incompatible types of chemicals and likely decomposition products

VII. Spill, Leak, and Disposal Procedures: includes step-by-step information

VIII. Special Protection Information: describes equipment needed for safe use

IX. Special Precautions and Comments: describes storage requirements and other notes

WARD'S has a pocket guide (Ward's Catalog No. 32 T 0002) that explains in greater detail how to use Material Safety Data Sheets.

## *Chemical Hazard Classes*

The hazards presented by any chemical can be grouped into the following categories. (It is important to keep in mind that a particular chemical may have more than one of these hazards.)

- **Flammable**
- **Corrosive**
- **Poisonous (toxic)**
- **Reactive**

A fifth category, those chemicals that do not possess the above properties, are termed "low hazard" materials. Water is an example of a low hazard material. Although these materials may not ordinarily represent a hazard, their presence in the lab requires that they be treated differently than they would be in a kitchen or backyard.

The following precautions should be used with all hazard classes:

- Require the use of safety goggles, gloves, and lab aprons.
- Minimize the amounts available in the lab (100 mL or less).
- Become familiar with first-aid measures for each chemical used.
- Become familiar with incompatibility issues for each chemical used.
- Emphasize how essential lab and storeroom cleanliness and personal hygiene are when dealing with hazardous materials.
- Keep hazardous chemicals in approved containers that are kept closed and stored away from sunlight and rapid temperature changes.
- Store chemicals of each hazard class away from those in other hazard classes.
- Keep a designated and locked storage cabinet for each hazard class.

## *FLAMMABLE*

### *Prevention and Control Measures*

- Store away from oxidizers and reactives.
- Keep containers closed when not in use.

- The ignition source is the easiest of the three components to remove. Check for the presence of lighted burners, sources of sparks (including static charge, friction, and electrical equipment), and hot objects such as hot plates or incandescent bulbs.
- Ground (electrically) all bulk metal containers when dispensing flammable liquids.
- Flammable vapors are usually heavier than air and can travel considerable distances before being diluted below ignitable concentrations.
- Ensure that there are class B fire extinguishers present in the laboratory and store room.
- Students should be drilled in EXACTLY what they must do if their clothes or hair catches fire. Practice "drop and roll" techniques. Both a safety shower and fire blankets should be available. Inform students that the shower is the best way to put out a fire on polyester clothing.
- Conduct a fire inspection with members of the local fire department at least once a year. Practice fire drills regularly.
- Provide adequate ventilation.
- Prepare for spills by having absorbent, vapor-reducing materials (available commercially) close at hand. Plan to have enough absorbent material to handle the maximum volume of flammable substances on hand.

### *Additional Protective Equipment*

- Nitrile or butyl rubber gloves
- Approved storage containers
- Face shield (recommended)
- Fire blanket
- Safety shower
- Fire extinguishers (Class B)

## *CORROSIVE*

### *Prevention and Control Measures*

- Always wear a face shield along with safety goggles when handling solutions of any corrosive material with concentrations greater than 1 mol/L.
- Have an eyewash station in close proximity.
- Wear the correct type of hand protection that will be impervious to the corrosive being handled. (Nitrile or butyl rubber gloves are generally recommended.)
- Provide adequate ventilation.
- Prepare for spills by having neutralizing reagents close at hand and in sufficient quantities for materials on hand.

### *Additional Protective Equipment*

- Nitrile or butyl rubber gloves
- Safety shower
- Face shield
- Eyewash stations
- Sleeve gauntlets

## POISONOUS (TOXIC)

### Prevention and Control Measures

- Treat all chemicals as toxic until proven otherwise. Above all, emphasize barriers, cleanliness, and avoidance of contact when handling any chemical.
- Wear protective equipment over exposed skin areas and eyes.
- Handle all contaminated glass and metal carefully. Remember that any sharp object can be a vehicle for introducing a toxic substance.
- Provide good ventilation. Use a chemical fume hood if possible.
- Recognize symptoms of overexposure and typical routes of introduction for each chemical used during a lab activity.
- Remember that the skin is not a good barrier to many toxic chemicals.
- Post the phone number of the nearest poison control center ON your phone.

### Additional Protective Equipment

- Container for sharp objects
- Chemical fume hood
- Particle face mask

## Chemical Storage

Never store chemicals alphabetically, as that greatly increases the risk of a violent reaction. Take these additional precautions.

1. Always lock the storeroom and all cabinets when not in use.
2. Do not allow students in the storeroom or preparation areas.
3. Avoid storing chemicals on the floor of the storeroom.
4. Do not store chemicals above eye level or on the top shelf in the storeroom.
5. Be sure shelf assemblies are firmly secured to walls.
6. Provide antiroll lips for all shelves.
7. Use shelving constructed of wood. Metal cabinets and shelves are easily corroded.
8. Avoid metal adjustable shelf supports and clips. They can corrode, causing shelves to collapse.
9. Store acids in their own locking storage cabinet.
10. Store flammables in their own locking storage cabinet.
11. Store poisons in their own locking storage cabinet.
12. Store oxidizers by classification, preferably in their own locking storage cabinets.

# Additional Resources

Your school district may have more information on safety issues. Some districts have a safety officer responsible for safety throughout their schools. Other possible sources for information include your state education agency, teachers' and science teachers' associations, and local colleges or universities.

The **American Chemical Society Health and Safety Service** will refer inquiries about health and safety to appropriate resources.

American Chemical Society (ACS)
1155 Sixteenth Street, N.W.
Washington, D.C. 20036
(202) 872-4511

## *Hazardous Materials Information Exchange (HMIX)*

Sponsored by the Federal Emergency Management Agency and the United States Department of Transportation, HMIX serves as a reliable on-line database. It can be accessed through an electronic bulletin board, and it provides information regarding instructional material and literature listings, hazardous materials, emergency procedures, and applicable laws and regulations.

HMIX can be accessed by a personal computer with a modem. Dial (312) 972-3275. The bulletin board is available 24 hours a day, seven days a week. The service is available free of charge. You pay only for the telephone call.

## *Safety Reference Works*


Gessner, G. H., ed. *Hawley's Condensed Chemical Dictionary* (11th ed.). New York: Van Nostrand Reinhold, 1987.

*A Guide to Information Sources Related to the Safety and Management of Laboratory Wastes from Secondary Schools.* New York State Environmental Facilities Corp., 1985.

Lefevre, M. J. *The First Aid Manual for Chemical Accidents.* Stroudsberg, Pa: Dowdwen, 1989.

Pipitone, D., ed. *Safe Storage of Laboratory Chemicals.* New York: John Wiley, 1984.

*Prudent Practices for Disposal of Chemicals from Laboratories.* Washington, D.C.: National Academy Press, 1983.

*Prudent Practices for Handling Hazardous Chemicals in Laboratories.* Washington, D.C.: National Academy Press, 1981.

Strauss, H., and M. Kaufman, eds. *Handbook for Chemical Technicians.* New York: McGraw-Hill, 1981.

*WARD'S MSDS Database and User's Guide.* WARD'S CD-ROM, Catalog Number 74T5070.

# *Release and Disposal of Organisms*

The ultimate responsibility falls on the teacher to ensure that each organism brought into the classroom receives adequate care during its stay, release, and final disposition.

The following information is provided to help you make informed decisions regarding the organisms used in the *Biotechnology* labs.

## *General Guidelines*

- Microbes—Bacteria, fungi, yeast, and growth media or materials that have come into contact with these organisms should not be discarded without decontamination (sterilization). See: "Safety With Microbes" in the *Laboratory Safety* section.

  *Bacillus licheniformis*—Destroy by autoclaving or disinfecting with household bleach (5% sodium hypochlorite).

  *Bacillus thuringiensis*—Destroy by autoclaving or disinfecting with household bleach (5% sodium hypochlorite).

  Oil-degrading microbes: *Penicillium* and *Pseudomonas*—Destroy by autoclaving or disinfecting with household bleach (5% sodium hypochlorite).

# Chemical Handling and Disposal

This information is furnished without warranty of any kind. Teachers should use it only as a supplement to other information they have and should make independent determinations of its suitability and completeness as it relates to their own district guidelines.

## HAZARDOUS CHEMICALS

Before using or disposing of any of the materials listed below, familiarize yourself with the safety and handling procedures and storage information listed under "Reagents and Storage" in the *Laboratory Safety* section. Also refer to individual reagent labels and Material Safety Data Sheets for further information about hazards and precautions.

### Container Labeling

Ensure that each container used by students in the laboratory is properly labeled with the following information:

- the name of the material and its concentration (if a solution)
- the names of individual components and their respective concentrations (if a mixture)
- the appropriate SIGNAL WORD
- a declarative statement of potential hazards
- immediate first-aid measures

Example:

> Lugol's Iodine Solution
> **WARNING: Poison if ingested. Irritant.**
> Do not ingest. Avoid skin and eye contact.
> Flush spills and splashes with water for 15 minutes; rinse mouth with water.
> Call your teacher immediately.

You should be aware of local, state, and federal regulations governing the disposal of hazardous materials. Contact a licensed Treatment, Storage, and Disposal (TSD) facility for disposal of large quantities of hazardous chemicals. Disposal protocols outlined below are ONLY for the substances (and quantities) specified.

Unless your school's drains are connected to a sanitary sewer system, no chemicals should ever be flushed down the drain. Never pour chemicals or reagents down the drain if you have a septic system. Even if you are connected to a sanitary sewer, do not pour any chemical down the drain unless you are certain it is safe and permitted.

The chemicals and reagents are classified by a color code to indicate hazard level. Remember to store chemicals of different hazard classes away from each other. YELLOW: Reactives, RED: Flammables, WHITE: Corrosives, BLUE: Toxics (poisons), GREEN: Low hazard for laboratory use.

## Disposal Method A—Inert Solid Wastes (Low Hazard)

Items identified by *DISPOSAL: METHOD A* can be considered to be a LOW HAZARD for laboratory handling and disposal. Avoid creating or breathing dust from these materials. If necessary, seal these materials in a bag or other suitable container. These materials generally have a GREEN storage code (general storage). It is recommended that you check local, state, and federal regulations to be certain that these articles may be disposed of in a sanitary landfill.

## Disposal Method B—Small Quantities of Liquid Wastes (Low Hazard)

Items identified by *DISPOSAL: METHOD B* can be considered to be a LOW HAZARD for laboratory handling and disposal. Wear the following PPE (personal protective equipment) when handling or disposing of these items: chemical safety goggles, apron, and polyethylene or nitrile gloves. Work in an area near an eyewash station. Use the following method to dispose of volumes of no more than 250 mL (unless specifically stated). Dilute the volume of waste material with 20 times as much tap water. Test pH with litmus or other indicator; if necessary, adjust pH to neutrality by adding small amounts of 1 M acid, base, or other reagent as required. (Exceptions are stated in the list below.) Place a beaker containing the diluted mixture in the sink, and run water to overflowing for 10 minutes, flushing to a known sanitary sewer. These materials generally have a GREEN storage code (general storage). It is recommended that you check local, state, and federal regulations to be certain that these articles may be disposed of in a sanitary landfill.

## Agarose—prepared (0.8%, 2%)

- **PPE:** Thermal hand protection, face shield, apron
- **PREPARATION:** Wear PPE. GEL CASTING—Unscrew cap 3/4 turn from tight. Melt agarose using a double-boiler, water bath, or microwave oven (use LOW power setting and heat at no more that 1 min intervals until complete melting occurs). Wear PPE when handling. Guard against superheating—superheated materials can be ejected from a container and cause burns. Exercise caution when GENTLY swirling bottles to mix between microwave heating.
- **STORAGE:** GREEN (general storage)
- **DISPOSAL:** LOW HAZARD for laboratory handling. METHOD A

## Bleach (5% sodium hypochlorite solution) NOT FOR STUDENT USE

- **CAUTION: Reactive material; strong oxidant.** Strong irritant. Avoid skin and eye contact; avoid vapor inhalation. Vapors are irritating to the upper respiratory tract; prolonged inhalation may cause edema. In case of contact, flush affected areas with water for 15 minutes, including under the eyelids; get medical attention if redness or irritation persists. If ingested, drink one or two glasses of water; contact a physician.
- **PPE:** Chemical safety goggles, nitrile gloves, apron. Working under a chemical fume hood is recommended.
- **STORAGE:** YELLOW (reactive material) Do not mix with acids or other oxidizers.
- **DISPOSAL:** METHOD B

## Calcium chloride solution (0.05 M)

- **CAUTION:** Mild irritant. Avoid contact with skin and eyes. Do not ingest.
- **PPE:** Chemical safety goggles, nitrile or polyethylene gloves, apron

- **PREPARATION:** Recommend using purchased solution. Wear PPE. Dissolve 5.55 g $CaCl_2$ in 75 mL of distilled water. Add additional distilled water to bring the final volume to 100 mL. Sterilize the solution by passing it through a prerinsed Swinnex water filter into an appropriate container or autoclave or steam sterilize at 121°C and 15 psi for 15–20 minutes.
- **STORAGE:** GREEN (general storage) Store in amber bottles; keep away from light.
- **DISPOSAL:** METHOD B

## *DNA (lambda)—digests*

- **STORAGE:** GREEN (general storage) Store frozen.
- **DISPOSAL:** METHOD A

## *DNA ligase (T4) (from T4-infected* E. coli*)*

- **STORAGE:** GREEN (general storage) Store frozen.
- **DISPOSAL:** METHOD A

## *Disinfectant solution (10% bleach)*

- **CAUTION:** Irritant. Avoid contact with skin and eyes. Do not ingest.
- **PPE:** Chemical safety goggles, nitrile or polyethylene gloves, apron
- **PREPARATION:** Add 100 mL of household bleach solution to 900 mL of distilled water.
- **STORAGE:** GREEN (general storage) Store in amber bottles; keep away from light.
- **DISPOSAL:** METHOD B

## *WARD'S DNA stain*

- **CAUTION:** Irritant. Avoid skin and eye contact; can stain skin and clothing. In case of contact, flush affected areas with water for 15 minutes, including under the eyelids; rinse mouth with water.
- **PPE:** Chemical safety goggles, polyethylene gloves, apron
- **STORAGE:** GREEN (general storage)
- **DISPOSAL:** METHOD B

## *N, N—dimethylformamide (DMF) NOT FOR STUDENT USE*

- **WARNING: Strong irritant/poison.** Harmful if inhaled, swallowed, or absorbed through the skin. Vapor or mist is irritating to eyes, mucous membranes, and upper respiratory tract. In case of contact, immediately flush eyes and skin with copious quantities of water for at least 15 minutes, including under the eyelids. If inhaled, remove to fresh air. DMF is a potent liver toxin. Avoid consumption of any alcoholic beverage for one week following inhalation exposure. Read MSDS before use.
- **PPE:** Chemical safety goggles, nitrile gloves, apron, fume hood
- **PREPARATION:** Wear PPE. Read MSDS before use. Use only under a fume hood, and avoid dusting conditions. In a sterile centrifuge tube, dissolve premeasured quantity (38 mg) of X-gal crystals in 2 mL DMF (N,N —dimethylformamide). Keep on ice until needed. Wash hands following use.
- **STORAGE:** BLUE (poison) Store away from oxidizers. Keep container tightly closed.
- **DISPOSAL:** Material is completely used up in preparation of *Luria agar with DMF, X-gal, and IPTG.*

## *Electrophoresis dye set*

Set contains 1.0 mL quantities of the following: Bromophenol blue, crystal violet, orange G, methyl green, xylene-cyanol, dye mixture.

- **CAUTION:** Irritant. Avoid skin and eye contact; can stain skin and clothing. In case of contact, flush affected areas with water for 15 minutes, including under the eyelids; rinse mouth with water.
- **PPE:** Chemical safety goggles, polyethylene gloves, apron
- **DISPOSAL:** METHOD B

## *Ethyl alcohol (70%)*

- **WARNING: Flammable liquid.** Avoid open flames, excessive heat, sparks, and other potential ignition sources; do not ingest; avoid eye contact; avoid prolonged skin contact. In case of contact, flush affected areas with water for 15 minutes, including under the eyelids; rinse mouth with water. Get prompt medical attention.
- **PPE:** Chemical safety goggles, nitrile or polyethylene gloves, apron

- **STORAGE:** RED (flammable liquid)
- **DISPOSAL:** Wear PPE. Dilute small volumes (less than 250 mL) in a ratio of one part solution to 20 parts water. Place a beaker of the diluted solution in the sink, and run water to overflowing, flushing to a sanitary sewer.

## *Ethyl alcohol (95%) NOT FOR STUDENT USE*

- **WARNING: Flammable liquid.** Avoid open flames, excessive heat, sparks, and other potential ignition sources; do not ingest; avoid eye contact; avoid prolonged skin contact. In case of contact, flush affected areas with water for 15 minutes, including under the eyelids; rinse mouth with water. Get prompt medical attention.
- **PPE:** Chemical safety goggles, nitrile or polyethylene gloves, apron
- **STORAGE:** RED (flammable liquid)
- **DISPOSAL:** Wear PPE. Dilute small volumes (less than 250 mL) in a ratio of one part solution to 20 parts water. Place a beaker of the diluted solution in a sink, and run water to overflowing, flushing to a sanitary sewer.

## *Fertilizer (microbiological growth enhancer)*

- **PPE:** Chemical safety goggles, nitrile or polyethylene gloves, apron. Avoid dusting conditions during application.
- **STORAGE:** GREEN (general storage)
- **DISPOSAL:** METHOD B

## *Glycerin NOT FOR STUDENT USE*

- **PPE:** Chemical safety goggles, polyethylene gloves, apron
- **STORAGE:** GREEN (general storage)
- **DISPOSAL:** METHOD B

## *Ice-nucleating protein/INP solution*

- **CAUTION:** Avoid creating dusting conditions.
- **STORAGE:** GREEN (general storage)
- **DISPOSAL:** METHOD A (solid ); METHOD B (solution)

## *Isopropyl-ß-D-Thiogalactopyrananoside (IPTG) solution NOT FOR STUDENT USE*

- **CAUTION: Strong irritant/poison.** Harmful if inhaled or swallowed; avoid skin and eye contact; do not ingest. In case of contact, immediately flush eyes and skin with a copious amount of water for at least 15 minutes, including under the eyelids. If inhaled, remove to fresh air. Read MSDS before use.
- **PPE:** Chemical safety goggles, nitrile gloves, apron, fume hood
- **PREPARATION:** Wear PPE. Read MSDS before use. Use only under a fume hood; avoid dusting conditions. In a vial, dissolve the premeasured quantity of IPTG (23.8 mg) in 1 mL of sterile distilled water. Wash hands following use.
- **STORAGE:** BLUE (poison) Store away from oxidizers. Keep container tightly closed.
- **DISPOSAL:** Material is completely used up in preparation of *Luria agar w/ X-gal, IPTG, and DMF.*

## *Ligation buffer (1X) (tris-KCl-EDTA solution)*

- **CAUTION:** Mild irritant. Avoid skin and eye contact; flush affected areas with water for 15 minutes, including under the eyelids; rinse mouth with water.
- **PPE:** Chemical safety goggles, nitrile or polyethylene gloves, apron
- **STORAGE:** GREEN (general storage)
- **DISPOSAL:** METHOD B

## *WARD'S Loading dye (bromophenol blue/glycerol solution)*

- **CAUTION:** Irritant. Avoid skin and eye contact; can stain skin and clothing. In case of contact, flush affected areas with water for 15 minutes, including under the eyelids; rinse mouth with water.

- **PPE:** Chemical safety goggles, polyethylene gloves, apron
- **STORAGE:** GREEN (general storage)
- **DISPOSAL:** METHOD B

## *Luria agar/broth*

- **STORAGE:** GREEN (general storage) Store under refrigeration (6°C) until needed.
- **DISPOSAL:** LOW HAZARD for laboratory handling when sterile. Autoclave or steam sterilize (121°C and 15 psi for 15–20 minutes) contaminated material. May also be chemically sterilized by applying a thin layer of 70% isopropyl alcohol to the surface of plates, waiting 20 minutes, taping plates, placing them in sealed plastic bags, and discarding them as inert solid waste.

## *Luria agar with X-gal/IPTG/DMF*

- **CAUTION:** Contains minute amounts of X-gal (3.5 mg/plate), IPTG (1.3 mg/plate), and DMF (0.18 mL/plate). Avoid skin contact with agar; do not ingest. DMF (a toxin) can be absorbed through skin.
- **PPE:** Polyethylene gloves
- **STORAGE:** BLUE (poison) Store under refrigeration (6°C) until needed.
- **DISPOSAL:** Autoclave or steam sterilize (121°C and 15 psi for 15–20 minutes) contaminated material. May also be chemically sterilized by applying a thin layer of 70% isopropyl alcohol to the surface of plates, waiting 20 minutes, taping plates, placing them in sealed plastic bags, and discarding them as inert solid waste. NOTE: Do not use bleach to decontaminate.

## *Luria agar with tetracycline*

- **STORAGE:** GREEN (general storage) Store under refrigeration (6°C) until needed.
- **DISPOSAL:** LOW HAZARD for laboratory handling when sterile. Autoclave or steam sterilize (121°C and 15 psi for 15–20 minutes) contaminated material. May also be chemically sterilized by applying a thin layer of household bleach (5% sodium hypochlorite) to surface of plates, waiting 20 minutes, taping plates, and discarding them as inert solid waste.

## *Methyl green/pyronin stain*

- **CAUTION:** Irritant. Avoid skin and eye contact; can stain skin and clothing. In case of contact, flush affected areas with water for 15 minutes, including under the eyelids; rinse mouth with water.
- **PPE:** Chemical safety goggles, polyethylene gloves, apron
- **STORAGE:** GREEN (general storage) Protect from light.
- **DISPOSAL:** METHOD B

## *Motor oil—virgin (refined)*

- **CAUTION: Irritant/combustible liquid.** Avoid skin and eye contact and open flames or ignition sources; do not ingest. If contact occurs, flush affected areas with water for 15 minutes; rinse mouth with water.
- **PPE:** Chemical safety goggles, polyethylene gloves, apron
- **STORAGE:** RED (combustible/flammable liquid)
- **DISPOSAL:** Dispose through approved sources—usually area automobile repair shops or others licensed to receive waste motor oil.

## *Mounting medium (Piccolyte II)*

- **CAUTION: Irritant/combustible liquid.** Avoid skin and eye contact and open flames or ignition sources; do not ingest. If contact occurs, flush affected areas with water for 15 minutes; rinse mouth with water.
- **PPE:** Chemical safety goggles, polyethylene gloves, apron
- **STORAGE:** RED (combustible liquid)
- **DISPOSAL:** Allow mixture to air-dry until solid, then discard.

## Onion root tip storage medium (83% denatured ethyl alcohol)

- **WARNING: Flammable liquid.** Avoid open flames, excessive heat, sparks and other potential ignition sources; do not ingest; avoid eye contact; avoid prolonged skin contact. In case of contact, flush affected areas with water for 15 minutes, including under the eyelids; rinse mouth with water. Get prompt medical attention.
- **PPE:** Chemical safety goggles, nitrile or polyethylene gloves, apron
- **STORAGE:** RED (flammable liquid)
- **DISPOSAL:** Wear PPE. Dilute small volumes (less than 50 mL) in a ratio of 1 part solution to 20 parts water. Place a beaker of the diluted solution in the sink, and run water to overflowing, flushing to sanitary sewer.

## Sodium dodecyl sulfate (10%)/sodium chloride (1.5%)—SDS solution

- **CAUTION:** Mild irritant. Avoid skin and eye contact. In case of contact, flush affected areas with water for 15 minutes, including under the eyelids; rinse mouth with water.
- **PPE:** Chemical safety goggles, nitrile or polyethylene gloves, apron
- **PREPARATION:** Wear PPE. Avoid dusting conditions. Add 10 g of sodium dodecyl sulfate and 1.5 g of sodium chloride to 50 mL of distilled water. Stir to dissolve. Add additional distilled water to bring the final volume to 100 mL.
- **STORAGE:** GREEN (general storage)
- **DISPOSAL:** METHOD B

## TBE running buffer (1X) (tris borate EDTA)

Each liter contains 0.089 mol of tris base, 0.089 mol of boric acid, and 0.0020 mol of EDTA disodium salt.

- **CAUTION:** Mild irritant. Avoid skin and eye contact. In case of contact, flush affected areas with water for 15 minutes, including under the eyelids; rinse mouth with water.
- **PPE:** Chemical safety goggles, nitrile or polyethylene gloves, apron
- **STORAGE:** GREEN (general storage)
- **DISPOSAL:** METHOD B

## Tryptic soy agar (slants)

- **PREPARATION:** Dissolve 40 g of tryptic soy agar dehydrated microbiological media in 1 L of distilled water. Bring the mixture to a boil, and dispense into bottles or tubes. Autoclave at 121°C and 15–18 psi for 15–20 minutes. Allow to cool. To dispense, melt in a microwave oven at 30 sec intervals (make sure caps are open to allow gas release) and dispense 15–20 mL into sterile petri plates. Allow to cool; refrigerate upside down to avoid condensation contamination.
- **STORAGE:** GREEN (general storage) Store under refrigeration (6°C) until needed.
- **DISPOSAL:** LOW HAZARD for laboratory handling when sterile. Autoclave or steam sterilize (121°C and 15 psi for 15–20 minutes) contaminated material. May also be chemically sterilized by applying a thin layer of household bleach (5% sodium hypochlorite) to the surface plates, waiting 20 minutes, taping plates, and discarding them as inert solid waste.

## X-gal (5-Bromo-4-chloro-3-indolyl-ß-D-galactopyranoside) NOT FOR STUDENT USE

- **CAUTION: Irritant/poison.** Harmful if inhaled or swallowed; avoid skin and eye contact; do not ingest. Read MSDS before use.
- **PPE:** Chemical safety goggles, nitrile gloves, apron, fume hood
- **PREPARATION:** Wear PPE. Read MSDS before use. Work under a fume hood; avoid dusting conditions. In a sterile centrifuge tube, dissolve premeasured quantity (38 mg) of X-gal crystals in 2 mL DMF (N,N —dimethylformamide). Keep on ice until needed.
- **STORAGE:** BLUE (poison) Store away from oxidizers.
- **DISPOSAL:** Material completely used up in preparation of *Luria agar w/ X-gal.*

# Master Materials List, by Category

This materials cross-reference guide was prepared by WARD'S Natural Science, the preferred science supplier for the Holt BioSources Lab Program published by Holt, Rinehart and Winston.

Materials are grouped into five categories: Biological Supplies, Chemicals and Media, Laboratory Equipment, Kits, and Miscellaneous. Each entry is listed alphabetically, followed by package size and the WARD'S catalog number. The second column gives the number of the lab in which the material is used. A list of the materials needed for each lab follows.

WARD'S also has available a convenient and easy computer software-ordering system specifically designed for use with the Holt BioSources Lab Program. The software-ordering system lists all required and supplemental materials needed for every lab. Click on the products you need, and the software automatically creates your shopping list, keeping track of the materials you ordered and their costs. The software-ordering system is available for both Macintosh and IBM-compatible computers. Call WARD'S at 1-800-962-2660 for your free copy.

To order, or for questions concerning the use of WARD'S materials, call toll-free 1-800-962-2660 or fax WARD'S at 1-800-635-8439.

WARD'S™ WARD'S Natural Science Establishment, Inc.
5100 W. Henrietta Road
P.O. Box 92912
Rochester, NY 14692-9012
Internet address: http://www.wardsci.com
E-mail address: customer_service@wardsci.com

## Biological Supplies

| Item | Lab |
|---|---|
| *Bacillis licheniformis,* culture (85 T 0204) | **D9** |
| *Bacillis thuringiensis,* culture (85 T 0203) | **D9** |
| DNA/RNA in Animal Cells, slide, each (93 T 2275) | **D1** |
| *Escherichia coli,* freeze-dried, culture (85 T 1666) | **D3** |
| *Escherichia coli* strain RRI, culture (85 T 1869) | **D4** |
| Lambda DNA *Eco*RI digest, 10 μg vial (85 T 1332) | **D6, D7** |
| Lambda DNA *Eco*RI, *Hin*dIII digest, 10 μg vial (85 T 1337) | **D6** |
| Lambda DNA *Hin*dIII digest, 10 μg vial (85 T 1331) | **D6, D7** |
| Lambda phage DNA, 50 μg vial (85 T 1330) | **D8** |
| Nucleic Acids I (SECT) MGP, slide, each (93 T 2322) | **D1** |
| Onion root tips, vial, (63 T 1212) | **D1** |
| *Penicillium* sp., freeze-dried, culture (85 T 7110) | **D11, D12** |
| Plasmid pBR322, 1.25 μg, culture (85 T 1300) | **D4** |
| Plasmid pUC8, 10 μg (85 T 1304) | **D3** |
| *Pseudomonas* sp., freeze-dried, culture (85 T 1704) | **D11, D12** |

## Chemicals and Media

| Item | Lab |
|---|---|
| Agarose, 2.0%, 200 mL btl, each (88 T 1210) | **D5** |
| Agarose, prepared 0.8%, 200 mL btl (88 T 1207) | **D6, D7, D8** |
| Agarose dye markers, set of 6 (36 T 5255) | **D5** |
| *Bam*HI restriction enzyme, 50 μL btl (85 T 1355) | **D8** |
| Bleach, chlorine, 1 pt btl (37 T 5554) | **D3, D4, D8, D9, D11, D12** |

| Item | Lab |
|---|---|
| Calcium chloride, 10 mL/vial, pkg of 10 (38 T 0937) | **D3, D4** |
| DNA stain for gel electrophoresis, 500 mL btl (38 T 9012) | **D6, D7, D8** |
| *Eco*RI restriction enzyme, 50 µL btl (85 T 1356) | **D8** |
| Ethyl alcohol, 95%, denatured, 500 mL btl (39 T 0277) | **D2** |
| *Hin*dIII restriction enzyme, 50 µL btl (85 T 1353) | **D8** |
| Ice-nucleating protein, each (36 T 5556) | **D10** |
| Isopropyl alcohol, 70% solution, 500 mL btl (39 T 4917) | **D9** |
| Loading dye, 6×, 5 mL btl (38 T 9115) | **D7, D8** |
| Luria agar, sterile, 200 mL, pkg of 5 (88 T 1205) | **D3** |
| Luria broth, tubes, pkg of 12 (88 T 0151) | **D3** |
| Methyl green-pyronin Y stain, each (38 T 7001) | **D1** |
| Nutrient agar bottles, pkg of 6 (88 T 1500) | **D4** |
| Nutrient broth, 15 mL/tube, pkg of 6 (88 T 0503) | **D4, D11** |
| NYSM agar plates, pkg of 6 (88 T 0931) | **D9** |
| NYSM medium, 100 mL btl (88 T 1506) | **D9** |
| NYSM medium, 500 mL btl (88 T 1505) | **D9** |
| Phenotype reagent set, each (750 T 0600) | **D3** |
| Piccolyte II, 120 mL btl (37 T 9530) | **D1** |
| SDS 10%/NaCl 1.5% solution, each (37 T 3002) | **D2** |
| T4 DNA ligase 2/buffer, 200 units, each (85 T 1380) | **D7** |
| Tetracyline-HCl, 250 mg capsules, pkg of 100 (21 T 2816) | **D4** |
| Tris-borate-EDTA (TBE) buffer, 10× sol., 250 mL btl (37 T 0620) | **D5, D6, D7, D8** |
| Tryptic soy agar, tubes, pkg of 12 (88 T 0817) | **D3** |
| Water, distilled, 1 gal btl (88 T 7005) | **D3, D5, D6, D7, D8, D10, D11, D12** |
| Water (local) | **D4** |
| Water, sterile hydrating solution, pkg of 4 (88 T 0502) | **D3** |

## *Laboratory Equipment*

| Item | Lab |
|---|---|
| Apron, disposable polyethylene, box of 100 (15 T 1050) | **D1, D2, D3, D4, D5, D6, D7, D8, D9, D10, D11, D12** |
| Aquarium air pump, Whisper 100, each (21 T 2982) | **D9** |
| Autoclave, 15 1/2 qt capacity, each (14 T 9001) | **D3** |
| Balance, triple-beam, each (15 T 6057) | **D12** |
| Beaker, low-form 250 mL Griffin, each (17 T 4040) | **D5, D6, D7, D8** |
| Beaker, low-form 600 mL Griffin, each (17 T 4060) | **D2, D3, D4, D7, D8** |
| Biohazard bag, 12 × 24 in., pkg of 100 (18 T 6905) | **D3, D9, D11** |
| Bunsen burner, std. nat. gas, each (15 T 0612) | **D3, D4** |
| Clamp, Hoffman screw-compressor, each (15 T 3910) | **D9** |
| Coverslips, 22 mm plastic, box of 100 (14 T 3555) | **D1** |
| Culture tube, 16 × 125 mm, pkg of 25 (18 T 7169) | **D3, D4** |
| Dispenser, economy with enlargeable spout, 15 mL, pkg of 12 (18 T 1553) | **D11** |
| Electrophoresis chamber, dual gel, each (36 T 5160) | **D5, D6, D7, D8** |
| Electrophoresis system, battery-powered, each (36 T 5164) | **D5, D6, D7, D8** |
| Erlenmeyer flask, graduated 1000 mL, each (17 T 2985) | **D9** |
| Forceps, dissecting, medium, each (14 T 1001) | **D1, D9** |
| Funnel, analytical 60-degree angle, each (18 T 1331) | **D2** |
| Gas lighter, flat-file, pkg of 10 (15 T 0683) | **D3, D4** |
| Gloves, disposable, medium, box of 100 (15 T 1071) | **D3, D4, D9, D11, D12** |
| Gloves, heat defier kelnit cotton, pair (15 T 1095) | **D3, D4** |
| Graduated cylinder, 10 mL PP, each (18 T 1705) | **D2, D9** |
| Graduated cylinder, 250 mL PP, each (18 T 1740) | **D5, D6, D7, D8** |
| Graduated cylinder, 50 mL PP, each (18 T 1720) | **D3** |
| Hot plate, 700 W single-burner, each (15 T 7999) | **D3, D4, D7** |
| Incubator, lab, each (15 T 0060) | **D3, D4, D11, D12** |
| Inoculating loop, disposable (14 T 0954) | **D4** |
| Inoculating loop, nichrome, each (14 T 0957) | **D2, D3, D9** |
| Microcentrifuge-tube rack, each (18 T 4205) | **D5, D6, D7, D8** |

| Item | Lab |
|---|---|
| Microcentrifuge tubes, 1.5 mL, pkg of 500 (18 T 1361) | **D3, D5, D6, D7, D8** |
| Micropipet, 1–5 µL student, pkg of 250 (15 T 2096) | **D7, D8** |
| Micropipet, 10 µL economy fixed-volume, 10 µL capacity, each (15 T 2101) | **D4, D5, D6, D7, D8** |
| Micropipet tips, 5–200 µL capacity, pkg of 200 (15 T 1715) | **D4, D5, D6, D7, D8** |
| Microplates, 96-well, pkg of 5 (18 T 1372) | **D10** |
| Microscope, the WARD'S scope, each (24 T 2310) | **D1, D2** |
| Microscope slide, qual. precleaned, pkg of 72 (14 T 3500) | **D1, D2** |
| Mini stirrer w/speed adjustment, each (15 T 1531) | **D9** |
| Mortar, porcelain, size 0, 50 mL, each (15 T 3334) | **D2** |
| Pestle, porcelain, size 0, each (15 T 3335) | **D2** |
| Petri dish, disposable 100 × 15 mm, pkg of 20 (18 T 7101) | **D3, D4** |
| pH paper, 1–14 range, vial of 100 (15 T 2558) | **D9** |
| Pipet, nonsterile 6 in., pkg of 500 (18 T 2971) | **D10** |
| Pipet, serological disp. plastic, pkg of 100 (18 T 7196) | **D3, D4** |
| Pipet, sterile graduated, 6 in., pkg of 500 (18 T 2972) | **D3, D4, D9, D11, D12** |
| Pipet bulbs, pkg of 10 (15 T 0511) | **D3, D4** |
| Ruler, 6 in., white vinylite, each (14 T 0810) | **D5, D6, D7, D8** |
| Safety goggles, SG34 regular, each (15 T 3046) | **D1, D2, D3, D4, D5, D6, D7, D8, D9, D10, D11, D12** |
| Scoop, laboratory, 6 1/2 in., each (15 T 4339) | **D11** |
| Spatula, broad 11 1/8 in., each (15 T 9853) | **D6, D7, D8** |
| Stoppers, No. 8 twist-it, bag of 16 (15 T 8408) | **D9** |
| Stoppers, black rubber, solid, size 0, lb (15 T 8460) | **D10** |
| Swab applicator, pkg of 100 (14 T 5502) | **D3, D4** |
| Syringe filters, sterile, pkg of 15 (18 T 1580) | **D9** |
| Syringe, disposable 10 cc, each (14 T 1617) | **D9** |
| Test-tube holder, foam floating (local) | **D3, D4** |
| Test-tube rack, 6-well LDPE, each (18 T 4231) | **D2, D3, D4, D10** |
| Test tube with rim, 13 × 100 mm Pyrex, each (17 T 0610) | **D9** |
| Test tube with rim, 15 × 125 mm Pyrex, each (17 T 0620) | **D2, D10** |
| Thermometer, lab, –20 to 110°C, each (15 T 1416) | **D3, D4, D7, D10** |
| Tubing, rigid plastic 11.5 in. × 3/16 in., each (18 T 7700) | **D9** |
| Tubing, vinyl 3 × 5/16 in., 10-ft roll (18 T 5081) | **D9** |
| Washing bottle, fine stream, 240 mL, pkg of 6 (18 T 4150) | **D3, D4, D8, D9, D11, D12** |

## *Miscellaneous*

| Item | Lab |
|---|---|
| Aluminum foil, 12 in. wide, 25-ft roll (15 T 1009) | **D10** |
| Bags, resealable zipper, 4 × 6 in., pkg of 10 (18 T 6921) | **D8** |
| Battery pack, 9 V alkaline, pkg of 5 (14 T 5420) | **D5, D6, D7, D8** |
| Bovine liver, 2 × 2 cm, (local) | **D2** |
| Calculator, slimline TI-1100+, each (27 T 3055) | **D7, D8, D10** |
| Cardboard strips, 1/2 × 2 in. (local) | **D10** |
| Cheesecloth, 5-yd pkg (15 T 0015) | **D2** |
| Density indicator strips, pkg of 25 (15 T 1996) | **D11, D12** |
| Filter paper strips, 5 × 50 mm, pkg of 12 (250 T 2836) | **D9** |
| Freezer, (local) | **D8, D10** |
| Graph paper, log (local) | **D8** |
| Ice, (local) | **D2, D3, D4, D7, D8** |
| Jar, clear polystyrene, 2 oz, 53 mm, each (18 T 1632) | **D11, D12** |
| Jar cap, white metal, 53 mm, each (17 T 2133) | **D11, D12** |
| Marker, black lab, each (15 T 3083) | **D3, D4, D8** |
| Motor oil, SAE 30, 4 oz btl (37 T 0005) | **D11, D12** |
| Nutrient fertilizer, 1 oz pkt (750 T 3501) | **D11, D12** |
| Paper towel, 100-sheet 2-ply roll, each (15 T 9844) | **D1, D3, D4, D8, D9, D11, D12** |
| Pen, black wax marker for glass, each (15 T 1155) | **D10, D11, D12** |
| Pencils, Ticonderoga No. 2, box of 12 (15 T 9816) | **D1** |
| Petroleum jelly, 1 oz pkg (15 T 9832) | **D9** |

| Item | Lab |
|---|---|
| Sand, fine white, 32 oz pkg (45 T 1983) | **D2** |
| Sand, fine white, 500 g bag (20 T 7423) | **D12** |
| Stapler, each (15 T 1955) | **D10** |
| Sticks, craft, pkg of 30 (15 T 9893) | **D1** |
| Stopwatch, digital, each (15 T 0512) | **D1, D4, D7, D10** |
| Tape, transparent w/dispenser, each (15 T 1959) | **D3, D4, D8** |
| Tray, w/cover 4 × 7 × 1 3/4 in., each (18 T 0031) | **D6, D7, D8** |

## *Kits*

| Item | Lab |
|---|---|
| DNA and RNA Staining Lab, each (36 T 1215) | **D1** |
| DNA Ligation Lab Activity Kit, each (36 T 5377) | **D7** |
| Electrophoresis Introduction, set (36 T 5166) | **D5** |
| Ice-Nucleating Bacteria Study Kit, each (85 T 3501) | **D10** |
| Introduction to Biotechnology DNA Extraction, each (36 T 6026) | **D2** |
| Microbes at Work: Antibiotic Production, each,(36 T 6056) | **D9** |
| Oil-Degrading Microbes Oil Spill, each (85 T 3503) | **D11, D12** |
| Phenotype Expression, AP Lab No. 6, each (88 T 8410) | **D3** |

## *Supplemental Materials*

| Item | Lab |
|---|---|
| Antibiotic Resistance Set, each (85 T 3960) | **D4** |
| Antibiotic Sensitivity Lab, each (88 T 8105) | **D4** |
| *Bacteria* (2nd Edition), VHS video, each (193 T 0267) | **D10, D11, D12** |
| Bacteria Types, classroom chart, each (33 T 0764) | **D10, D11, D12** |
| *Bacterial Transformation*, VHS video, each (193 T 2141) | **D3, D4** |
| DNA Extraction Concept Study Kit, each (88 T 8109) | **D2** |
| DNA Fingerprinting Electrophoresis, set (36 T 5167) | **D7, D8** |
| DNA Fragment Analysis/Plasmid DNA, kit (88 T 8500) | **D6** |
| *DNA Replication and Mitosis*, VHS video, each (193 T 6452) | **D1** |
| *DNA Sequencing*, VHS video, each (193 T 2143) | **D1, D7** |
| DNA Spooling in Onion Demo Lab, kit (36 T 6021) | **D2** |
| *DNA Technology—Awesome Skill*, book, each (32 T 8076) | **D2, D6, D7, D8** |
| Drosophila Chromosome Extraction, kit (36 T 6010) | **D2** |
| *Electrophoresis Manual*, each (32 T 0807) | **D5, D6, D7, D8** |
| *Electrophoresis Simulation*, Mac, each (74 T 4056) | **D5, D7, D8** |
| Extracting Bacterial DNA Kit, each (88 T 8112) | **D2** |
| Fermentation Manual, set (32 T 0812) | **D9** |
| Fermentation Vessel Set, each (14 T 0300) | **D9** |
| Gel Electrophoresis Simulation Kit, each (36 T 1604) | **D5, D8** |
| *Gel Electrophoresis Training*, VHS video, each (193 T 5300) | **D5, D6, D7, D8** |
| Growth of Bacillus Cultures, kit (85 T 3511) | **D9** |
| Introduction to Fermentation Process, kit (85 T 3510) | **D9** |
| *Introduction to Bacteria*, book, each (32 T 1706) | **D3, D10, D11, D12** |
| *Microbes: Bacteria/Fungi*, laserdisc, each (196 T 0035) | **D10, D11, D12** |
| *New Look at Bacteria*, VHS video, each (193 T 2012) | **D10, D11, D12** |
| Plasmid Mini-Prep and Restriction, kit (88 T 8504) | **D7, D8** |
| *Restriction Enzyme Digestion*, VHS video, each (193 T 2140) | **D8** |
| Restriction Mapping of DNA Gel Kit, each (88 T 8503) | **D8** |
| Transformation of E. coli Kit, each (88 T 8230) | **D3** |

# Master Materials List, by Lab

## D1 Laboratory Techniques: Staining DNA and RNA

| *Required Materials* | *Quantity Needed* |
|---|---|
| Apron, disposable polyethylene | 1 per student |
| Coverslips, 22 mm Plastic | 1 per team |
| DNA/RNA in Animal Cells Slide | 1 per team |
| Forceps, dissecting, medium | 1 per team |
| Methyl green-pyronin Y stain | 1 mL per team |
| Microscope slide, qual. precleaned | 1 per team |
| Microscope, the WARD'S Scope | 1 per team |
| Nucleic Acids I (SECT) MGP, slide | 1 per team |
| Onion root tips | 1 per team |
| Paper towel, 100-sheet, 2-ply roll | 1 sheet per team |
| Pencil, Ticonderoga No. 2 | 1 per team |
| Piccolyte II | 1 mL per team |
| Safety goggles, SG34 regular | 1 per student |
| Sticks, popsicle | 1 per team |
| Stopwatch, digital | 1 per team |
| **Alternative Activity** | |
| DNA and RNA Staining Lab | 1 per 30 students |

## D2 Laboratory Techniques: Extracting DNA

| *Required Materials* | *Quantity Needed* |
|---|---|
| SDS 10%/NaCl 1.5% solution | 10 mL per team |
| Apron, disposable polyethylene | 1 per student |
| Beaker, low-form 600 mL Griffin | 1 per team |
| Bovine liver, 2 × 2 cm | 1 piece per team |
| Cheesecloth, 12 × 12 cm pieces | 2 per team |
| Ethyl alcohol, 95%, denatured | 4 mL per team |
| Funnel, analytical 60-degree angle | 1 per team |
| Graduated cylinder, 10 mL PP | 1 per team |
| Ice | 8–10 cubes per team |
| Inoculating loop, nichrome | 1 per team |
| Mortar, porcelain, size 0, 50 mL | 1 per team |
| Pestle, porcelain, size 0 | 1 per team |
| Safety goggles, SG34 regular | 1 per student |
| Sand, fine white | 5 grams per team |
| Test-tube rack, 6-well LDPE | 1 per team |
| Test tube with rim, 15 × 125 mm Pyrex | 1 per team |
| **Alternative Activity** | |
| Introduction to Biotechnology DNA Extraction | 1 per 30 students |

## D3 Laboratory Techniques: Genetic Transformation of Bacteria

| *Required Materials* | *Quantity Needed* |
|---|---|
| Apron, disposable polyethylene | 1 per student |
| Autoclave, 15-1/2 qt capacity | 1 per class |
| Beaker, low-form 600 mL Griffin | 2 per team |
| Biohazard bag, 12 × 24 in. | 1 per team |
| Bleach, chlorine | 1 per class |
| Bunsen burner, std. nat. gas | 1 per class |
| Culture tube, 16 × 125 mm | 3 per team |
| Calcium chloride | 500 µL per team |
| Eppendorf digital pipette | 1 per team |
| *Escherichia coli*, freeze-dried culture | 1 per class |
| Gas lighter, flat file | 1 per class |
| Gloves, disposable, medium | 2 gloves per student |
| Gloves, heat defier kelnit cotton | 1 per class |
| Graduated cylinder, 50 mL PP | 1 per class |
| Hot plate, 700 W single-burner | 1 per team |
| Ice | 8–10 cubes per team |
| Incubator, lab | 1 per team |
| Inoculating loop, nichrome | 2 per team |
| Luria agar, sterile, 200 mL bottle | 1 per team |
| Luria broth | 5 mL per team |
| Marker, black lab | 1 per team |
| Microcentrifuge tube | 1 per team |
| Paper towel, 100-sheet 2-ply roll | 5 sheets per team |
| Petri dish, disposable 100 × 15 mm | 4 per team |
| Phenotype Reagent Set | 1 per team |
| Pipet bulb | 1 per team |
| Pipet, serological disp. plastic | 6 per team |
| Pipet, sterile graduated | 4 per team |
| Plasmid, pUC8, rehydrated | 0.01 mL per team |
| Safety goggles, SG34 regular | 1 per student |
| Swab applicator | 4 per team |
| Tape, transparent w/dispenser | 1 per team |
| Test-tube rack | 1 per team |
| Thermometer, lab –20 to 110°C | 1 per team |
| Tryptic soy agar, tubes | 5 slants per class |
| Tube holder, foam floating | 1 per team |
| Washing bottle, fine stream | 2–3 per class |
| Water, sterile hydrating solution | 1 bottle per class |
| **Alternative Activity** | |
| Phenotype Expression, AP Lab #6 | 1 per 24 students |

## D4 Experimental Design: Genetic Transformation—Antibiotic Resistance

| *Required Materials* | *Quantity Needed* |
|---|---|
| Apron, disposable polyethylene | 1 per student |
| Beaker, low-form 600 mL Griffin | 2 per team |
| Bleach, chlorine | 1 per team |
| Bunsen burner, std. nat. gas | 1 per team |
| Calcium chloride | 0.50 mL per team |
| Culture tube, 16 × 125 mm | 2 per team |

## MASTER MATERIALS LIST continued

*Escherichia coli* strain RRI . . . . . . . . . . . . 1 per class
Gas lighter, flat file . . . . . . . . . . . . . . . . . . 1 per team
Gloves, disposable, medium . . . . . . . . . 2 per student
Gloves, heat defier kelnit cotton. . . . . 1 pair per class
Hot plate, 700 W single-burner . . . . . . . 1 per team
Ice . . . . . . . . . . . . . . . . . . . . . . 8–10 cubes per team
Incubator, lab . . . . . . . . . . . . . . . . . . . . . 1 per class
Inoculating loop, disposable . . . . . . . . . . 2 per team
Marker, black lab . . . . . . . . . . . . . . . . . . 1 per team
Nutrient agar bottles . . . . . . . . . . . . . . . . 1 per team
Nutrient broth . . . . . . . . . . . . . . . . . . 5 mL per team
Paper towel, 100-sheet
  2-ply roll . . . . . . . . . . . . . . . . . 5 sheets per team
Petri dish, disposable 100 × 15 mm. . . . 4 per team
Pipet bulb . . . . . . . . . . . . . . . . . . . . . . . . 2 per team
Pipet, serological disp. plastic . . . . . . . . . 6 per team
Pipet, sterile graduated . . . . . . . . . . . . . . 4 per team
Plasmid pBR322, rehydrated . . . . . . 10 µL per team
Safety goggles, SG34 regular . . . . . . . . 1 per student
Stopwatch, digital . . . . . . . . . . . . . . . . . . 1 per team
Swab applicator . . . . . . . . . . . . . . . . . . . 4 per team
Tape, transparent w/dispenser . . . . . . . . . 1 per class
Test-tube rack . . . . . . . . . . . . . . . . . . . . . 1 per team
Tetracyline-HCl, 250 mg capsules . . . . . . 1 per class
Thermometer, lab –20 to 110°C . . . . . . . . 1 per team
Tube holder, foam floating. . . . . . . . . . . . 1 per team
Washing bottle, fine stream . . . . . . . . . . 2–3 per class
Water . . . . . . . . . . . . . . . . . . . . . . . 500 mL per team

### D5 Laboratory Techniques: Introduction to Agarose Gel Electrophoresis

| *Required Materials* | *Quantity Needed* |
|---|---|
| Agarose dye markers | 1 per class |
| Agarose gel, 2.0%, precast | 1 per team |
| Apron, disposable polyethylene | 1 per student |
| Battery pack/5, 9-V alkaline | 1 per team |
| Beaker, low-form 250 mL Griffin | 1 per team |
| Electrophoresis system, battery-powered | 1 per team |
| Graduated cylinder, 250 mL PP | 1 per team |
| Microcentrifuge-tube rack | 1 per team |
| Microfuge tubes, 1.5 mL | 6 per team |
| Micropipet, economy fixed-volume | 1 per team |
| Micropipet tips, 5–200 µL | 6 per team |
| Ruler, 6 in. white vinylite | 1 per team |
| Safety goggles, SG34 regular | 1 per student |
| Tris-borate-EDTA buffer, 10× sol. | 20 mL per team |
| Water, distilled | 180 mL per team |
| **Alternative Activity** | |
| Electrophoresis Introduction | 1 per class |

### D6 Laboratory Techniques: DNA Fragment Analysis

| *Required Materials* | *Quantity Needed* |
|---|---|
| Agarose, prepared 0.8% | 15 mL per 2 teams |
| Apron, disposable polyethylene | 1 per student |
| Battery pack/5, 9-V alkaline | 1 per team |
| Beaker, low-form 250 mL Griffin | 1 per team |
| DNA stain for gel electrophoresis | 100 mL per team |
| Electrophoresis system, battery-powered | 1 per 2 teams |
| Graduated cylinder, 250 mL PP | 1 per team |
| Lambda DNA *Eco*RI digest | 10 µL per team |
| Lambda DNA *Eco*RI, *Hin*dIII digest | 10 µL per team |
| Lambda DNA *Hin*dIII digest | 10 µL per team |
| Microcentrifuge-tube rack | 1 per team |
| Microfuge tubes, 1.5 mL | 3 per team |
| Micropipet, economy 10 µL fixed-volume | 1 per team |
| Micropipet tips, 5–200 µL | 3 per team |
| Ruler, 6 in. white vinylite | 1 per team |
| Safety goggles, SG34 regular | 1 per student |
| Spatula, lab 11 1/8 in. | 1 per team |
| Tray, w/cover 4 × 7 × 1 3/4 in. | 1 per team |
| Tris-borate-EDTA buffer, 10× sol. | 20 mL per team |
| Water, distilled | 280 mL per team |

### D7 Laboratory Techniques: DNA Ligation

| *Required Materials* | *Quantity Needed* |
|---|---|
| Agarose, prepared 0.8% | 15 mL per 2 teams |
| Apron, disposable polyethylene | 1 per student |
| Battery pack/5, 9-V alkaline | 1 per team |
| Beaker, low-form 250 mL Griffin | 1 per team |
| Beaker, low-form 600 mL Griffin | 3 per team |
| Calculator, slimline TI-1100+ | 1 per team |
| DNA stain for gel electrophoresis | 100 mL per team |
| Electrophoresis system, battery-powered | 1 per 2 teams |
| Graduated cylinder, 250 mL PP | 1 per team |
| Hot plate, 700 W single-burner | 1 per team |
| Ice | 8–10 cubes per team |
| Lambda DNA *Eco*RI digest | 20 µL per team |
| Lambda DNA *Hin*dIII digest | 10 µL per team |
| Loading dye, 6× | 2 µL per team |
| Microcentrifuge-tube rack | 1 per team |
| Microcentrifuge tubes, 1.5 mL | 3 per team |
| Micropipet, 10 µL economy fixed-volume | 1 per team |
| Micropipet tips, 5–200 µL | 3 per team |

Micropipets, 1–5μL student . . . . . . . . . . . 3 per team
Ruler, 6 in. white vinylite . . . . . . . . . . . . . 1 per team
Safety goggles, SG34 regular . . . . . . . . 1 per student
Spatula, lab 11 1/8 in. . . . . . . . . . . . . . . . . 1 per team
Stopwatch, digital . . . . . . . . . . . . . . . . . . . 1 per team
T4 DNA ligase 2/buffer, 200 units . . . . 3 μL per team
Thermometer, lab –20 to 110°C . . . . . . . 1 per team
Tray, w/cover 4 × 7 × 1 3/4 in. . . . . . . . . 1 per team
Tris-borate-EDTA buffer,
10× sol. . . . . . . . . . . . . . . . . . . . 20 mL per team
Water, distilled . . . . . . . . . . . . . . . . 280 mL per team

**Alternative Activity**

DNA Ligation Lab Activity Kit . . . . . . . . . 1 per class

## D8 Experimental Design: Comparing DNA Samples

| *Required Materials* | *Quantity Needed* |
|---|---|
| Agarose, prepared 0.8% | 15 mL per 2 teams |
| Apron, disposable polyethylene | 1 per student |
| Bags, resealable zipper, 4 × 6 in. | 1 per team |
| *Bam*HI restriction enzyme | 2 μL per team |
| Beaker, low-form 250 mL Griffin | 1 per team |
| Beaker, low-form 600 mL Griffin | 2 per team |
| Bleach, chlorine | 1 per class |
| Calculator, slimline TI-1100+ | 1 per team |
| DNA stain for gel electrophoresis | 200 mL per team |
| *Eco*RI restriction enzyme | 4 μL per team |
| Electrophoresis system, battery-powered | 1 per 2 teams |
| Freezer | 1 per class |
| Graduated cylinder, 250 mL PP | 1 per team |
| Graph paper, log | 2 sheets per team |
| *Hin*dIII restriction enzyme | 2 μL per team |
| Ice | 8–10 cubes per team |
| Lambda phage DNA | 12 μL per team |
| Loading dye, 6× | 3 μL per team |
| Marker, black lab | 1 per team |
| Microcentrifuge-tube rack | 1 per team |
| Microcentrifuge tubes, 1.5 mL | 3 per team |
| Micropipet, 10 μL economy fixed-volume | 1 per team |
| Micropipet tips, disposable 5–200 μL capacity | 3 per team |
| Micropipets, 1–5 μL student | 3 per team |
| Paper towel, 100-sheet 2-ply roll | 3 sheets per team |
| Power supply, 3-cell | 1 per team |
| Ruler, 6 in. white vinylite | 1 per team |
| Safety goggles, SG34 regular | 1 per student |
| Spatula, lab 11 1/8 in. | 1 per team |
| Tray, w/cover 4 × 7 × 1 3/4 in. | 1 per class |
| Tris-borate-EDTA buffer, 10× sol. | 20 mL per team |
| Washing bottle, fine stream | 2–3 per class |
| Water, Distilled | 100 mL per team |

## D9 Laboratory Techniques: Introduction to Fermentation

| *Required Materials* | *Quantity Needed* |
|---|---|
| Apron, disposable polyethylene | 1 per student |
| Aquarium air pump, whisper 100 | 1 per team |
| *Bacillis licheniformis*, culture | 1 per team |
| *Bacillis thuringiensis*, culture | 1 per team |
| Biohazard bag, 12 × 24 in. | 1 per class |
| Bleach, chlorine | 1 per class |
| Clamp, Hoffman screw-compressor | 2 per team |
| Erlenmeyer flask, graduated 1000 mL | 1 per team |
| Filter paper strips, 5 × 50 mm | 12 per team |
| Forceps, dissecting, medium | 1 per team |
| Gloves, disposable, medium | 2 per student |
| Graduated cylinder, 10 mL PP | 1 per team |
| Inoculating loop, nichrome | 2 per team |
| Isopropyl alcohol, 70% solution | 1 per class |
| Mini stirrer w/speed adjustment | 1 per team |
| NYSM agar plates | 3 per team |
| NYSM medium, 100 mL btl. | 1 per team |
| NYSM medium, 500 mL btl. | 1 per team |
| Petroleum jelly | 1 per team |
| pH paper, 1–14 range | 3 strips per team |
| Pipet, sterile graduated, 6 in. | 1 per team |
| Safety goggles, SG34 regular | 1 per student |
| Stoppers, #8 twist-it | 1 per team |
| Syringe filters, sterile | 2 per team |
| Syringe, disposable 10 cc | 1 per team |
| Test tube with rim, 13 × 100 mm Pyrex. | 1 per team |
| Tubing, rigid plastic 11.5 in. × 3/16 in. | 3 per team |
| Tubing, vinyl 3 in. × 5/16 in. | 4 ft per team |
| Washing bottle, fine stream | 2–3 per class |
| **Alternative Activity** | |
| Microbes at Work: Antibiotic Prod. | 1 per team |

## D10 Laboratory Techniques: Ice-Nucleating Bacteria

| *Required Materials* | *Quantity Needed* |
|---|---|
| Aluminum foil, 12 in. wide roll | 1 ft per team |
| Apron, disposable polyethylene | 1 per student |
| Calculator, slimline TI-1100+ | 1 per team |
| Cardboard strips, 1/2 × 2 in. | 2 per team |
| Freezer | 1 per class |

Ice-nucleating protein . . . . . . . 4 granules per team
Microplates, 96-well . . . . . . . . . . . . . . . . . 1 per team
Pen, black wax marker for glass . . . . . . . . 1 per team
Pipet, 6 in. nonsterile . . . . . . . . . . . . . . . . 4 per team
Safety goggles, SG34 regular . . . . . . . . 1 per student
Stapler . . . . . . . . . . . . . . . . . . . . . . . . . . . 1 per team
Stopper, black rubber, solid, size 0 . . . . . . 1 per team
Stopwatch, digital . . . . . . . . . . . . . . . . . . . 1 per team
Test-tube rack, 6-well LDPE. . . . . . . . . . . 1 per team
Test Tube with rim, 15 × 125 mm
Pyrex . . . . . . . . . . . . . . . . . . . . . . . . . 3 per team
Thermometer, lab –20 to 110°C . . . . . . . 2 per team
Water, distilled . . . . . . . . . . . . . . . . . 15 mL per team

**Alternative Activity**

Ice Nucleating Bacteria Study Kit . . . . 1 per 6 teams

## D11 Laboratory Techniques: Oil-Degrading Microbes

| *Required Materials* | *Quantity Needed* |
|---|---|
| Apron, disposable polyethylene | 1 per student |
| Biohazard bag, 12 × 24 in. | 1 per class |
| Bleach, chlorine, 1 pt btl | 1 per class |
| Density indicator strips | 3 per team |
| Dispenser, economy (for oil) | 1 per team |
| Gloves, disposable, medium | 2 per student |
| Incubator, lab. | 1 per class |
| Jar cap, white metal 53 mm | 3 per team |
| Jars, clear polystyrene, 2 oz, 53 mm. | 3 per team |
| Motor oil, SAE 30 | 2 mL per team |
| Nutrient broth, 15 mL/tube | 2 tubes per team |
| Nutrient fertilizer | 1.5 g per team |
| Paper towel, 100-sheet 2-ply roll | 5 sheets per team |
| Pen, black wax marker for glass | 1 per team |
| *Penicillium* sp. freeze dried | 1 per team |
| Pipet, sterile 6 in. | 2 per team |
| *Pseudomonas* sp. freeze dried. | 1 per team |
| Safety goggles, SG34 regular | 1 per student |
| Scoop, laboratory | 1 per team |
| Washing bottle, fine stream (for bleach) | 2–3 per class |
| Water, distilled | 90 mL per team |
| **Alternative Activity** | |
| Oil-Degrading Microbes Oil Spill, kit | 1 per 24 students |

## D12 Experimental Design: Can Oil-Degrading Microbes Save the Bay?

| *Required Materials* | *Quantity Needed* |
|---|---|
| Apron, disposable polyethylene | 1 per student |
| Balance, triple beam | 1 per team |
| Bleach, chlorine, 1 pt btl | 1 btl per class |
| Density indicator strips | 6 per team |
| Gloves, disposable, medium | 2 per student |
| Incubator, lab. | 1 per class |
| Jar caps, white metal 53 mm | 6 per team |
| Jars, clear polystyrene, 2 oz, 53 mm. | 6 per team |
| Motor oil, SAE 30 | 10 mL per team |
| Nutrient fertilizer | 1.5 g per team |
| Pen black wax marker for glass | 1 per team |
| *Penicillium* sp. freeze dried. | 1 per team |
| Pipet, sterile graduated 6 in.. | 2 per team |
| *Pseudomonas* sp. freeze dried | 1 per team |
| Safety goggles, SG34 regular | 1 per student |
| Sand, fine white | 90 g per team |
| Washing bottle, fine stream | 2–3 per class |
| Water, distilled | 180 mL per team |
| **Alternative Activity** | |
| Oil-Degrading Micorbes Oil Spill, kit | 1 per 24 students |

# Laboratory Assessment

## *As a teacher, only you know the best assessment methods to apply to student lab work in your classes.*

After each lab, you may want students to prepare a lab report. A traditional lab report usually includes at least the following components:

- **title**
- **summary paragraph** describing the purpose and procedure
- **data tables** and **observations** that are organized and comprehensive
- **answers** to the Analysis and Conclusions questions

You may want students to use *Gowin's Vee.* Or, you may wish to use any of the other rubrics and checklists described below.

## *Teaching Resources CD-ROM Provides Customizable Rubrics, Checklists, and Gowin's Vees*

The **Holt BioSources Teaching Resources CD-ROM** includes fully editable *scoring rubrics* and *classroom-management checklists* to expand your assessment options. These customizable assessment tools can provide a means to objective assessment or serve as a good starting point for designing your own specialized assessment tools. Laboratory-related rubrics and checklists include the following:

- Introduction to Scoring Rubrics
- Student-Designed Experiments Rubric
- Informal Assessment Direct Observations Checklist
- Evaluating Laboratory Work
- Scoring rubrics for *Quick Labs, Inquiry Skills Development* labs, *Laboratory Techniques* and *Experimental Design* labs, and *Biotechnology* labs

*Using Gowin's Vee in the Lab* contains complete instructions for using Gowin's Vee, for those who wish to experience the benefits of this powerful teaching tool.

- Vee Form Instructions—Teacher's Notes
- Teaching Students to Use the Vee
- Vee Form (a blank Vee form that students can use with any lab)
- Assessing Student-Constructed Vee Reports

The **Holt BioSources Teaching Resources CD-ROM** also includes the following worksheets, which can help you emphasize the importance of safe lab behavior:

- Student Safety Contract
- Safety Quiz

# Using the Laboratory Techniques and Experimental Design Labs

### This book contains two types of laboratory activities.

- **Laboratory Techniques** labs teach your students laboratory skills and procedures in the context of a real-world biological career.
- **Experimental Design** labs require your students to apply these skills to design their own experiments as they take on the role of an employee at a scientific consulting firm.

### By using the Laboratory Techniques and Experimental Design labs, your students will do more than just get the answer. They will learn to:

- use techniques used by working biologists
- explore practical skills needed for successful achievement in the real world
- manage and allocate resources
- communicate effectively
- find information
- work as a team member
- select appropriate technology

## Laboratory Techniques Labs

Each *Laboratory Techniques* lab includes a detailed, step-by-step procedure similar to many traditional labs. Unlike *Quick Labs* and *Inquiry Skills Development* labs, Laboratory Techniques labs are placed in the context of real-life occupational scenarios. Students learn practical lab skills, procedures, and fundamental concepts of biology while they discover the role of biology in various careers.

### The parts of a Laboratory Techniques lab are described below.

- *Skills* identifies the techniques to be learned or practiced.
- *Objectives* describes the expected accomplishments.
- *Materials* is a list of the items required to perform the lab.
- *Purpose* describes the real-life setting.
- *Background* provides additional information needed for the lab.
- *Procedure* is a step-by-step explanation for doing the lab, including safety precautions and instructions for cleanup and disposal.
- *Analysis* questions ask students to examine the techniques and skills they used in the lab and to interpret their findings.
- *Conclusions* requires students to form opinions based on their observations and the data they collected.
- *Extensions* invites students to find out more about topics related to the subject of the lab.

## *Using Experimental Design Labs*

Each *Experimental Design* lab invites students to join BioLogical Resources, Inc., a biological consulting firm. As employees of the firm, students are given a business letter from a "client" that outlines a problem, and a memo from their supervisor that offers suggestions and guidance. Based on the problem, they must create a plan for their procedure, select supplies and equipment, and conduct the work to earn a profit for the firm.

**To ensure best results and minimal safety hazards, do not perform any *Experimental Design* lab without first performing its prerequisite *Laboratory Techniques* lab, if there is one indicated at the beginning of the lab.**

### *The parts of an Experimental Design lab are described below.*

- *Prerequisites* identifies the Laboratory Techniques lab that introduces procedures used.
- *Review* lists concepts students will need to know for the lab.
- *Business letter* poses a problem for students to solve.
- *Memorandum* is a note from the supervisor directing the student to perform the work. The memorandum also contains the following:
    - *Proposal Checklist,* which spells out what students are to do before the lab and which must be approved by you
    - *Report Procedures,* which spells out what students should include in their final report
    - *Required Precautions,* which lists the safety procedures students must follow
    - *Disposal Methods,* which provides instructions for the handling and removal of all materials used
- *Materials and Costs* includes everything students should need to do the lab plus some nonessential items, forcing students to think carefully and to request only what they need.

### *Experimental Design labs require an approach different from most traditional labs.*

If your students are new to working independently, they may initially feel uncomfortable when left on their own to propose a procedure for a self-directed lab. Be sure to take the following steps to help boost their confidence:

- **Perform the Laboratory Techniques** lab prerequisites, if there are any. Most Experimental Design labs are preceded by at least one prerequisite Laboratory Techniques lab.
- **Completely discuss the Laboratory Techniques lab prerequisites** with your students before they begin an Experimental Design lab.
- **Suggest that students use materials and equipment similar to those in the Laboratory Techniques** lab if they are unsure about how to begin.
- **Remind students to concentrate on what they need to know and on how they will measure it** so that they can remain focused.

- **Point out that it is more important to understand** what is going on in the lab than it is to perform the lab with excellent technique but no understanding.
- **Provide leading questions for students to consider** as they make their plans. Questions in the *Teacher's Notes and Sample Solutions* section that follows will help you underscore the applicability of techniques learned earlier.
- **Hold a question-and-answer session** before students begin the lab.
- **Hint (don't tell) that some of the items listed may be unnecessary.**
- **Remind students to check the Memorandum** in the Experimental Design lab for requirements for their proposal and final report.
- **Be thorough in helping students to develop their proposal,** as this will form the basis of their procedure. Each proposal should include:
    - the question(s) to be answered
    - the procedure to be used
    - at least one detailed data table
    - a list of proposed materials and costs
- **Emphasize that students cannot begin their procedure until you have approved their proposal and signed their *Proposal Checklist.***
- **Students can prepare their own invoice or use the *Materials and Costs* list of their lab as an invoice.** Give students the following tips for preparing an invoice:
    - An invoice should be generated from an itemized list of the costs incurred to perform the work requested by a client. Students should show the actual amount of time taken and the materials used, regardless of what they projected in their proposal.
    - An invoice should group itemized costs in meaningful categories, such as *Facilities and Equipment Use* and *Labor and Consumables.*
    - If you "fine" students for safety violations, have them subtract the fine when computing the *Subtotal* of their costs. Be sure that students notice the effect of the fine on the *Total Amount Due*—it will reduce the amount of profit made on the job.
    - To remain in business, the firm must show a profit. Students can compute the *Profit Margin,* which is the difference between the cost of the work and the amount the client is charged, by multiplying the *Subtotal* by a percentage, such as 0.25 (25 percent) or 0.30 (30 percent). Have students add the *Subtotal* and *Profit Margin* to compute the *Total Amount Due.*

### *Teacher's Notes and Sample Solutions ensure a safe, successful lab.*

The pages that follow contain Student Objectives, Preparation Notes, Procedural Notes, and Sample Solutions for each of the Experimental Design labs.

BIOTECHNOLOGY

# D4 Experimental Design: Genetic Transformation—Antibiotic Resistance

## Purpose

Students will genetically transform bacteria to test a plasmid carrying a gene for tetracycline resistance.

## INSTRUCTIONAL GOALS

### Student Objectives

- *Prepare* bacterial cells for genetic transformation.
- *Insert* a plasmid into a bacterial cell.
- *Grow* bacteria on agar dishes.
- *Analyze* bacterial growth on agar dishes and *interpret* the results.

## PREPARATION NOTES

**Time Required:** 2 or 3 class periods

### Pre-Lab Discussion

Review the procedure used in **Biotechnology D3—Laboratory Techniques: Genetic Transformation of Bacteria,** and answer any questions students may have. Also discuss with students the need for safety and aseptic technique in all labs that involve bacteria. Ask students the following questions to guide their thinking:

- What is a plasmid? *(Bacteria normally have two types of DNA. In addition to a main chromosome, they have a circular DNA molecule called a plasmid. Plasmids contain a relatively small genetic sequence that usually gives the cell a survival advantage. Plasmids are also self-replicating.)*
- What is the supposed survival advantage of the test plasmid? *(resistance to the antibiotic tetracycline)*
- What is genetic transformation? *(the process of changing an organism by transferring genetic material from another organism)*
- How do scientists verify that a plasmid carries a particular genetic trait? *(They isolate the plasmid, insert it in a bacterium without the trait, and then screen for evidence of genetic transformation.)*
- What is/are the independent variable(s) in this experiment? *(In essence, students will be testing for two different independent variables, the presence of tetracycline and the presence of the plasmid.)*
- Explain how you could set up controls for each independent variable. *(Set up four cultures so that bacteria with the plasmid and bacteria without the plasmid are each grown on agar with tetracycline and agar without tetracycline.)*

### Preparation Tips

- Always use aseptic or sterile technique when preparing stock solutions. Store each stock solution according to the directions on its package.
- The WARD'S Transformation of *E. coli* Kit contains all of the materials necessary for this activity. To prepare the solutions and cultures for this activity, follow the instructions enclosed with the kit.
- Prepare a dilute household bleach solution by adding 100 mL household bleach to 900 mL water, and pour it into squirt bottles labeled "disinfectant solution."
- Prepare liquid Luria broth, liquid agar, 0.5 M $CaCl_2$, and tetracycline mix according to the directions on the packages. Store tetracycline in the freezer until ready for use.
- Reserve half of the prepared agar for preparing petri dishes without tetracycline. Then prepare +T agar by adding 0.25 mL of prepared tetracycline to each 200 mL bottle of agar. Cap the bottle and swirl gently to mix.
- Prepare two −T petri dishes and two +T dishes for each student group. Flame the mouth of the bottle between dishes to avoid contamination.
- Rehydrate the *E. coli* RRI according to directions on the package, and transfer the bacteria to tryptic soy agar slants. Incubate these cultures at 37°C for 24 hours.
- Rehydrate the plasmid by adding 1 mL sterile water to lyophilized pBR322. Allow to stand undisturbed for one hour or rehydrate the day before it is needed and refrigerate.
- Set up ice baths and a 42°C hot-water bath for students ahead of time.
- Prepare plastic-foam tube holders by making tube-sized holes in a flat piece of plastic foam just small enough to float in the hot-water bath.
- Provide separate containers for collecting broken glass, unused nutrient broth, unused bacteria, unused plasmid, cotton swabs, micropipetter tips, inoculating loops, pipets, petri dishes, and other contaminated materials.

### Disposal

- Wear gloves to decontaminate spills and broken glass by covering them with paper towels and soaking the towels with a dilute household bleach solution. Let this stand for 30 minutes. Use tongs to collect all waste materials; then place them in a biohazard bag and seal the bag. Wash your hands with antibacterial soap after this procedure.
- Used slides should be placed in a household bleach solution for 24 hours.
- All contaminated materials should be autoclaved.

- All surfaces should be cleaned and decontaminated at the conclusion of each lab.
- Solid trash that has been contaminated with bacteria must be collected in a separate and specially marked disposal bag. Package all sharp instruments in separate metal containers for disposal. No pipets should protrude from the disposal bag. It is recommended that disposal bags (following autoclaving) be placed in an outer sealed container, such as a plastic bucket with a lid.
- Absorb liquids and gels with paper towels to minimize leakage.
- Contaminated materials that are to be decontaminated away from the laboratory must be placed in a durable, leakproof container that is closed prior to its removal from the laboratory.

## PROCEDURAL NOTES

### Safety Precautions

- Discuss all safety symbols and precautions with students.
- Remind students to wear oven mitts when handling hot objects.
- Remind students to use a pipet pump and not their mouths for pipetting.

### Procedural Tips

- If time limitations prevent completion of the procedure in one lab period, stop after the nutrient broth is added in step 11, leave it at room temperature for 24 hours, and then proceed with the next step.
- Remind students to use aseptic technique throughout the laboratory.
- When pouring the bacterial solution onto the agar, students should try to keep the solution in the center of the agar to prevent it from seeping down the sides of the dish.

## SAMPLE SOLUTION

### Question

Does the test plasmid carry a gene for tetracycline resistance?

### Proposed Procedure

1. Put on safety goggles, lab apron, and disposable gloves. Spray the lab table surface with the disinfectant solution, and wipe it with a paper towel.
2. Using a nontoxic permanent marker, label the lids of two 15 mL plastic tubes with the initials of everyone in the lab group. Label one lid "plasmid" and the second "no plasmid." Place the unopened tubes in an ice bath. Keep the tubes, bacteria, and plasmid on ice unless instructed otherwise.
3. Using a 1 mL sterile pipet, transfer 0.25 mL of ice-cold 0.5 M calcium chloride ($CaCl_2$) into each of two 15 mL tubes.
4. Use separate disposable inoculating loops to transfer several colonies from the stock culture of *E. coli* into each tube. Vigorously tap the inoculating loop against the wall of each tube to dislodge the cell mass. Then use a sterile pipet to gently mix the bacteria in the $CaCl_2$ solution, making sure the bacteria are suspended and that no cell mass is left on the side of the tube.
5. After each transfer, replace the tube lid and place the tube on ice.
6. Use a 1 mL sterile pipet to add 0.01 mL of the test plasmid to the plasmid tube only. Very gently tap the tube with your finger to mix the plasmid with the cell suspension, and return the tube to the ice bath.
7. Leave both tubes on ice for an additional 20 minutes.
8. During the 20 minutes, label four petri dishes as follows:
   Dish 1 plasmid, +T
   Dish 2 plasmid, −T
   Dish 3 no plasmid, +T
   Dish 4 no plasmid, −T
9. After 20 minutes, heat-shock the tubes by inserting them into a plastic-foam tube holder and transferring the tubes from the ice bath to the 42°C hot-water bath for 60 seconds.
10. Remove the tubes after 60 seconds and immediately place them back in the ice bath for 2 minutes, then remove them and allow them to return to room temperature.
11. Using separate graduated pipets, transfer 2.5 mL of nutrient broth into each tube. Gently tap the tubes to mix them. Incubate the tubes at 37°C for 30 minutes, or leave them overnight at room temperature.
12. Use a 1 mL plastic pipet to place 0.25 mL from the tube labeled "plasmid" onto Dish 1. Use a sterile cotton swab to spread the bacterial solution evenly over the agar.
13. Repeat step 12 to add bacteria with the plasmid to Dish 2. Repeat step 12 to add bacteria without the plasmid to Dish 3 and Dish 4. Use a new sterile pipet and sterile cotton swab for each dish.
14. Allow the petri dishes to sit for 10 minutes or until the agar has completely soaked up the liquid.
15. Seal each dish with transparent tape, invert the dishes, and then stack them together. Tape all

four dishes together and place them in the area designated for incubation. Incubate the dishes at 37°C for 24 to 48 hours.

16. After 24 to 48 hours, untape the stack of four dishes. Do not untape or open the individual dishes. Observe each dish, and record the description and number of colonies in a data table. The growth of fuzzy white colonies indicates the presence of bacteria.
17. Be sure to place all items in the correct disposal area. Clean up according to aseptic technique and the instructor's directions.

## Materials (per lab group)

- safety goggles
- disposable gloves
- lab aprons
- oven mitts
- incubator
- clock or watch with second hand
- two 600 mL beakers
- thermometer
- ice
- six 1 mL sterile plastic serological pipets
- hot plate
- 10µL of plasmid pBR322
- stock culture of *E. coli* RRI
- 5 mL of nutrient broth
- 2 petri dishes with nutrient agar −T (without tetracycline)
- 2 petri dishes with nutrient agar +T (with tetracycline)
- 0.50 mL of calcium chloride solution
- two 15 mL plastic tubes w/ lids
- test-tube rack
- 4 sterile cotton swabs
- 4 sterile graduated pipets
- 2 disposable inoculating loops
- plastic-foam tube holder
- pipet bulb
- disinfectant solution
- 3 paper towels
- nontoxic permanent marker
- 2 m of transparent tape

## Estimated Cost of Materials

$2672.41, including a 30 percent profit (based on two students per group, three days of facility use, and three hours of labor). Refer to *Using Experimental Design Labs* to see how to prepare an invoice.

## Results

Students should observe growth in all petri dishes except Dish 3, the one with tetracycline but no plasmid. Colonies of bacteria should grow in the absence of tetracycline, both with and without the plasmid. Colonies of bacteria should grow in the presence of tetracycline only when the plasmid has been inserted. Students should show the results in a data table similar to the one below:

**Sample Data Table**

| Dish | Plasmid (+ or −) | Tetracycline (+ or −) | Growth (yes or no) |
|---|---|---|---|
| Dish 1 (plasmid, +T) | + | + | yes |
| Dish 2 (plasmid, −T) | + | − | yes |
| Dish 3 (no plasmid, +T) | − | + | no |
| Dish 4 (no plasmid, −T) | − | − | yes |

## Conclusions

Students should conclude that, while the presence of tetracycline would normally prevent the growth of the *E. coli* bacteria, when the plasmid has been inserted, colonies of *E. coli* do grow in the presence of tetracycline. This indicates that the test plasmid does indeed contain a gene for tetracycline resistance.

# BIOTECHNOLOGY D8 *Experimental Design: Comparing DNA Samples*

## Purpose

Students will apply their technique in gel electrophoresis to compare the DNA of two subjects.

## Student Objectives

- *Compare* two DNA samples using the process of gel electrophoresis.

## PREPARATION NOTES

**Time Required:** two 50-minute lab periods

### Pre-Lab Discussion

Review the procedure from **Biotechnology D5—Laboratory Techniques: Introduction to Gel Electrophoresis** and **Biotechnology D6—Laboratory Techniques: DNA Fragment Analysis,** and answer any questions students may have. Prior to the lab, discuss with students the concept of gel electrophoresis—how and why it works. Discuss with students how to analyze a gel. You may do this by placing a gel on an overhead projector and showing students how to recognize the bands and calculate $R_f$ values. Model the proper use of all equipment for the student prior to the lab. Ask students the following question to guide their thinking:

What is $R_f$? *(It is the ratio of the distance a substance moves through a separation medium to the distance a reference substance, in this case the loading dye, moves through the medium. It is a measure of the relative mobility of a substance in a separation medium.)*

## Preparation Tips

- Purchase 0.125 μg/μL Lambda DNA. You may wish to purchase the DNA pre-cut by restriction enzymes.
- Cut the DNA in each microtube according to the following instructions:
  1. Prepare the following microtubes for each student group: a microtube labeled "Ms. Coleman" with 2 μL *Eco*RI; a microtube labeled "Ms. Wilson" with 2 μL *Bam*HI; and a microtube labeled "Marker Sample" with 2 μL *Hin*dIII.
  2. Place all microtubes in an ice bath and keep them there until they are ready to be put into the hot-water bath.
  3. Using a new 1–5 μL micropipet for each microtube, add 2 μL of the corresponding restriction buffer to each of the microtubes. Place the buffer on the side of the tube.
  4. Use a 10 μL fixed-volume micropipet to add 10 μL Lambda DNA to the side of each of the microtubes. Gently tap each microtube on your lab table until the solutions are thoroughly mixed. Use a new 5 to 200 μL micropipet tip for each tube.
  5. Place the microtubes in a 37°C water bath for 50–60 minutes. Then remove the microtubes from the water bath, and immediately put them into a freezer.
- You may wish to vary the experiment by preparing the DNA samples so that both are cut with the same enzyme, producing identical fragment lengths.
- Always keep microtubes with enzymes on ice. Microtubes come in a variety of colors. Students find it easier to keep microtubes separate if the tubes are different colors.
- Each restriction enzyme comes with the appropriate 10× restriction buffer. Keep the buffers on ice, and add only the corresponding 10× restriction buffer to each of the microtubes.
- Purchase 0.8% agarose ready to use or prepare it by adding 1.6 g of agarose to 200 mL of 1× TBE buffer. If the buffer comes as a 10× solution, dilute it by adding one part buffer to nine parts distilled water. Stir to suspend the agarose, and then heat it on a hot plate or in a microwave oven until dissolved. Cover and keep in a 65°C water bath until it is ready for use (makes 200 mL). Unused agarose solution can refrigerated and stored for several weeks.
- Purchase loading dye ready to use or prepare it by dissolving 0.25 g bromophenol blue, 0.25 g xylene cyanol, 50.0 g sucrose, and 1.0 mL 1M Tris (pH 8) in 60 mL deionized or distilled water. Dilute with deionized or distilled water to make a total volume of 100 mL.
- Prepare the gels according to the following instructions:
  1. Place a gel comb in the grooves of a gel-casting tray. Make sure that the comb does not touch the bottom of the tray. If it does, get another comb.
  2. Using a hot mitt, pour 15 mL of melted 0.8% agarose into a 50 mL graduated cylinder, and then pour the agarose into the gel-casting tray. Make sure that the gel spreads evenly throughout the tray. The thickness of the gel should not exceed 3mm.
  3. Let the gel cool (about 20–30 minutes) until the agarose solidifies.
  4. Carefully remove the gel comb by pulling it straight up. Do not wiggle the comb. This may tear the gel. If the gel does not come up easily, pour a little TBE buffer on the comb area. Now loosen the end of the gel tray and, holding a zipper-lock plastic bag open, carefully slide the gel into the bag. Do not let the gel fold over on itself; it will break. The gel can be stored this way for 1–2 days.
- Provide a separate container for the disposal of the following: broken glass, gel, unused 1× TBE buffer, unused loading dye, unused DNA stain, and disinfectant solution microtubes, micropipets, and micropipetter tips.

## Disposal

- Follow the instructions on the WARD'S Material Safety and Data Sheets (MSDS) for the disposal of solutions and dried gels.
- Paper towels and disposable gloves may be placed in a trash can.

# PROCEDURAL NOTES

## Safety Precautions

- Discuss all safety symbols and precautions with students.
- Caution students to be very careful when working with chemicals such as stains. Remind them to notify you immediately of any chemical spills. Also caution them to never taste, touch, or smell any substance or to bring it close to their eyes, unless specifically directed to do so.

## Procedural Tip

- Have students use the "gel handler" spatula to remove the gel from the staining tray. Wrap the gel in clear plastic wrap and store it in the refrigerator. Gel bands will retain the stain for about one month before they start to fade. Fading is due to the oxidation of DNA because DNA is not fixed in the gel matrix.
- When using a battery-operated electrophoresis chamber, it may take more than 2 hours for the DNA to separate. If you use three batteries instead of five, the DNA can be separated overnight.

## SAMPLE SOLUTION

### Question

Can either Ms. Berg or Ms. Wilson be identified as the twin of Ms. Coleman?

### Proposed Procedure

1. Put on safety goggles and a lab apron.
2. Place 1 μL of loading dye onto the side of each microtube. Use a new 1–5 μL micropipet for each tube, and do not touch the bottom of the tube. Tap on the lab bench to mix thoroughly; do not shake the microtubes. If the materials in the microtube are frozen, warm them in your hand until they thaw.
3. Carefully remove the gel from the plastic bag, and place it in the electrophoresis chamber of an electrophoresis apparatus so that the wells are closest to the black wire, or negative electrode.
4. Set a micropipetter to 10 μL, and place a new tip on the end. Open the microtube labeled "Ms. Coleman" and remove 10 μL.
5. Carefully dispense 10 μL of the material in the microtube into the well of lane 1. To do this, lean over the gel with both elbows on the lab table. Carefully lower the micropipetter tip into the opening of the well. Be careful that you do not jab the micropipetter tip through the bottom of the well. Depress the plunger to release the solution into the well.
6. Using a new pipetter tip for each tube, repeat steps 4 and 5 for each of the remaining microtubes. Use lane 2 for the microtube labeled "Ms. Wilson" and lane 3 for the microtube labeled "Marker Sample."
7. Very slowly fill the gel box with 1× TBE buffer so that the buffer covers the gel by approximately 2 mm.
8. Close the gel chamber, wipe off any spills, and connect it to a battery pack containing five 9V batteries.
9. Run your gel until the purple line reaches approximately 5 mm from the end of your gel, disconnect the battery pack, and remove the gel with a spatula. Store the gel in a plastic bag overnight prior to staining.
10. Carefully place the gel (wells up) in a staining tray. Pour DNA stain over the gel until it is completely covered.
11. Cover and label the staining tray, and allow the stain to sit for about two hours.
12. When staining is complete, pour off the stain into the sink drain and flush with water. Be careful that the gel does not slip out or break.
13. To destain, cover the gel with distilled water. Do not pour the water directly over the gel. Instead, pour it on the side of the staining tray. Let it sit for 18–24 hours.
14. Use a metric ruler to measure the distance in millimeters from the well in each lane to the tracking dye and DNA bands in each lane. Construct a data table for each DNA sample. Record the migration distance for each fragment in each lane in the appropriate data table. Then use the following equation to calculate the $R_f$ for each fragment of each sample:

$$R_f = \frac{\text{distance in mm that the DNA fragment migrated}}{\text{distance in mm from well to tracking dye}}$$

   Record these $R_f$ values in the appropriate data tables.
15. Enter the known fragment lengths of the marker sample in the marker sample data table, then calculate and record the log of each fragment length. Construct a graph plotting the log of each fragment length against $R_f$. Use this graph to find and record the logs of the fragment lengths for the unknown DNA samples. Calculate and record the fragment lengths for each fragment in each of the unknown samples. Compare and analyze the resulting data.

### Materials (per lab group)

- safety goggles
- lab aprons
- gel-casting tray with gel in plastic bag
- microtube rack
- 10 μl fixed-volume micropipetter
- staining tray
- battery-powered gel electrophoresis system
- freezer
- metric ruler
- calculator
- gel spatula
- three 1–5 μL micropipets
- 3 microtubes with cut DNA samples
- three 5–200 μL micropipetter tips

BIOTECHNOLOGY D8 *continued*

- 3 μL of loading dye
- 200 mL of DNA stain
- 200 mL of 1× TBE buffer
- 100 mL of distilled water
- bottle of disinfectant solution
- 3 paper towels
- 1 sheet of log graph paper

## Estimated Cost of the Job

$1,789.58, including a 30 percent profit (based on two students per group, two hours of labor, and two days of facility use). Refer to *Using Experimental Design Labs* to see how to prepare an invoice.

## Results

**Sample Data Tables**

[data tables below]

## Conclusions

Students' data should support their conclusions. Students should determine by the analysis of the gel that because the fragments of Ms. Berg do not match those of Ms. Coleman, the two cannot be twins.

**Standard Curve for the Lambda DNA/*Hin*dIII Marker Sample**

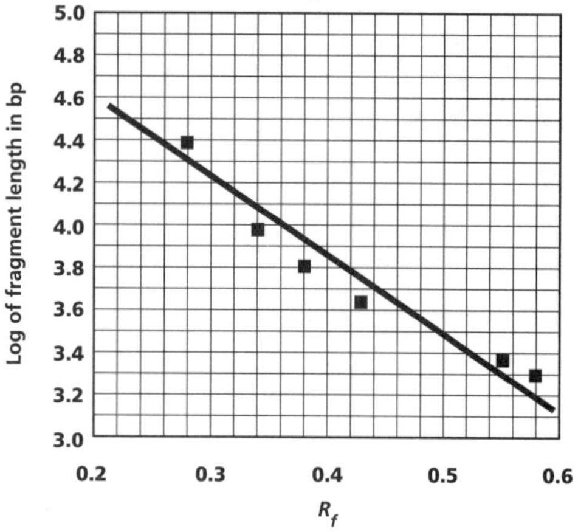


**DNA Marker Sample Data Table**

| Fragment number | Fragment length (bp) | Log | Migration distance (mm) | $R_f$ |
|---|---|---|---|---|
| 1 | 23,130 | 4.364 | 17 | 0.28 |
| 2 | 9,416 | 3.974 | 20 | 0.34 |
| 3 | 6,557 | 3.817 | 23 | 0.38 |
| 4 | 4,361 | 3.640 | 26 | 0.43 |
| 5 | 2,322 | 3.366 | 33 | 0.55 |
| 6 | 2,027 | 3.307 | 35 | 0.58 |
| 7 | 564 | 2.751 | not imaged | —— |
| 8 | 125 | 2.097 | not imaged | —— |

**Unknown DNA Sample Data Table**

| Ms. Coleman | | | | | Ms. Berg | | | | |
|---|---|---|---|---|---|---|---|---|---|
| Fragment number | Fragment length (bp) | Log of fragment length | Migration distance (mm) | $R_f$ | Fragment number | Fragment length | Log of fragment length | Migration distance | $R_f$ |
| 1 | 21,226 | 4.32 | 14.0 | 0.23 | 1 | 16,841 | 4.23 | 14.0 | 0.23 |
| 2 | 7,421 | 3.87 | 20.0 | 0.33 | 2 | 7,233 | 3.86 | 22.5 | 0.37 |
| 3 | 5,804 | 3.76 | 23.0 | 0.38 | 3 | 6,770 | 3.83 | 23.0 | 0.38 |
| 4 | 5,643 | 3.75 | 23.5 | 0.39 | 4 | 6,527 | 3.81 | 23.5 | 0.39 |
| 5 | 4,878 | 3.69 | 24.0 | 0.40 | 5 | 5,626 | 3.75 | 24.0 | 0.40 |
| 6 | 3,530 | 3.55 | 29.0 | 0.48 | 6 | 5,505 | 3.74 | 29.0 | 0.48 |

BIOTECHNOLOGY

# D12 Experimental Design: Can Oil-Degrading Microbes Save the Bay?

## Purpose

Students will test the role of microorganisms as a tool for environmental cleanup.

## Student Objectives

- *Design* an experiment to test the effectiveness of various treatments using microorganisms and microbe nutrients to break down refined oil.
- *Compare* the effectiveness of these different treatments in breaking down refined oil.
- *Recognize* the limitations of the tested oil-cleanup method.

## PREPARATION NOTES

**Time Required:** one 50-minute period for setup, plus 15 to 20 minutes a day for 4 days

## Pre-Lab Discussion

Review the procedure in **Biotechnology D11—Laboratory Techniques: Oil-Degrading Microbes,** and answer any questions students may have. Review aseptic technique used in previous labs. Discuss with students the effects of an oil spill on a marine environment. Ask students the following questions to guide their thinking:

- What is the difference between *Pseudomonas* and *Penicillium?* (*Pseudomonas is a bacterium and Penicillium is a fungus.*)
- How does a microbe density strip show the amount of growth? *(As microbes break down oil, the population density of the microbes increases*

*along with the turbidity, or cloudiness, of the solution. The greater the turbidity of the solution, the darker it gets, and the harder it becomes to see the lighter bars of the density strip through the solution. Density is read by observing the lightest color bar visible through the solution.)*

- Why is fertilizer used to help microorganisms break down oil? *(When fertilizer is added to an oil spill, it helps confine the oil to one area. Also, the fertilizer is an easier source of nutrients for the indigenous microorganisms to break down. Fertilizer stimulates the growth of microorganisms, which then consume the oil.)*
- What is Ms. Childs looking for in a treatment method? *(efficiency, low cost, low environmental impact, and appearance)*
- What kind of qualitative observations are important for answering Ms. Child's concerns? *(estimated density of microbes, appearance of water sample)*

## Preparation Tips

- Purchase live cultures of *Penicillium* and *Pseudomonas,* or follow these procedures to rehydrate freeze-dried cultures of bacteria and fungi:
  1. Lift the lid or plastic cover and remove the inner crystal. Using sterile technique, untwist the cap or cryovial.
  2. Use a sterile pipet to add 0.5 mL of nutrient broth to the pellet in the cryovial.
  3. Allow 60 seconds for the bacterial pellet to soften (10 minutes for the fungal pellet); then mix by drawing the suspension up and down through the pipet 8–10 times.
  4. If the kit has been purchased, use a sterile cotton swab and streak one slant tube with the *Pseudomonas.* Repeat with the second tube using a fresh cotton swab and *Penicillium.* These tubes can be stored for future use.
  5. Transfer approximately one-third of the remaining suspension to each of the three broth tubes. Incubate the tubes at 30°C for 3–4 days or until the tubes become cloudy.
  6. Autoclave all instruments used at 121°C and 15 psi for 20 minutes. Dispose of the materials in a biohazard bag.
- Prepare six water samples for each lab group by combining 30 mL of distilled water, 15 g of sand, 20 drops of refined oil, 1.25 mL from the prepared *Pseudomonas* culture, and 1.25 mL from the prepared *Penicillium* culture in each of six small jars with lids. When transferring cultures, use a sterile pipet each time.
- The indicator bars of the density indicator strips may show up more clearly if culture tubes are used instead of jars. Furthermore, students should be advised not to confuse the turbidity caused by fertilizer with the turbidity changes that occur as microbes degrade the oil.
- Prepare a 5 percent household-bleach disinfectant solution by adding 50 mL household bleach to 950 mL water. Pour the disinfectant into squirt bottles labeled "disinfectant solution."
- Provide separate containers for collecting broken glass and contaminated materials.

## Disposal

- Wear gloves to decontaminate spills and broken glass by covering them with paper towels and soaking the towels with a dilute household-bleach solution. Let this stand for 30 minutes. Then, use tongs to collect all waste materials, place them in a biohazard bag, and seal the bag. Wash your hands with antibacterial soap after this procedure.
- Used pipets and jars should be placed in a dilute solution of household bleach for 24 hours.
- All reusable materials that come in contact with bacteria or fungi should be autoclaved.
- All surfaces should be cleaned and decontaminated at the conclusion of each lab.
- Disinfectant solutions, staining solutions, and ethanol can be stored for future use.
- Solid trash that has been contaminated by bacteria or fungi must be collected in a separate and specially marked disposal bag. Package all sharp instruments in separate metal containers for disposal. No pipets should protrude from the disposal bag. It is recommended that disposal bags (following autoclaving) be placed in a sealed container, such as a plastic bucket with a lid. Liquids and gels must be absorbed by paper towel to minimize the risk of leakage.
- Contaminated materials not autoclaved in the laboratory must be placed in a durable, leakproof container that is closed prior to removal from the laboratory.

# PROCEDURAL NOTES

## Safety Precautions

- Discuss all safety symbols and precautions with students.
- Remind students to use a pipet pump and not their mouths for pipetting.

## Procedural Tip

Remind students to sterilize their work area before and after this lab.

## SAMPLE SOLUTION

### Question

What is the most effective treatment for degrading oil with microbes?

### Proposed Procedure

1. Put on safety goggles, disposable gloves, and a lab apron.
2. Pick up all materials. Label six water samples Control, Fertilizer, *Pseudomonas, Penicillium, Pseudomonas* with fertilizer, and *Penicillium* with fertilizer.
3. Using a sterile pipet, transfer 1.25 mL from the *Pseudomonas* culture into each of the jars marked *Pseudomonas* and *Pseudomonas* with fertilizer. Using a fresh pipet, transfer 1.25 mL from the *Penicillium* culture into each of the jars marked *Penicillium* and *Penicillium* with fertilizer.
4. Sprinkle 0.5 g of fertilizer into each of the jars marked Fertilizer, *Penicillium* with fertilizer, and *Pseudomonas* with fertilizer.
5. Carefully place a density indicator strip on each of the six jars so that the darkest bar is at the bottom. Close the jars.
6. Loosely cap and incubate the jars at 30°C.
7. Observe the jars daily for four more days and write observations in a data table. Look for any signs of oil degradation, such as the formation of tiny oil droplets, breakup of the oil layer into smaller fragments, or changes in texture. Note the color of the oil.

### Materials (per lab group)

- safety goggles
- disposable gloves
- lab aprons
- incubator
- balance
- 2 sterile graduated pipets
- water samples
- *Pseudomonas* culture
- *Penicillium* culture
- 1.5 g of nutrient fertilizer
- 6 density (turbidity) indicator strips
- wax pencil

### Estimated Cost of the Job

$3,029.20, including a 30 percent profit (based on two students per group, five days of facility use, and two total hours of labor). Refer to *Using Experimental Design Labs* to see how to prepare an invoice.

### Results

The signs of oil degradation should start to become evident after 1 or 2 days of incubation. Students should observe microbial growth over the oil layer and changes in the physical characteristics of the oil such as color and appearance. Students should note that as degradation proceeds, the initial continuous oil layer will have broken up and turned into finer and finer droplets. If a wet mount is made of these droplets, students will observe oil degrading bacteria massed around them. The *Penicillium* will result in a matted growth of cells on the surface of the oil. On the other hand, *Pseudomonas* will break up the oil into tiny droplets.

Students should observe an increase in turbidity (increased cell density) in the samples treated with oil-degrading bacteria over time. The addition of fertilizer should promote microbial growth and, therefore, increase turbidity. Fertilizer alone will promote microbial growth of any resident microlife. Degradation should continue until either oxygen (these microbes are aerobes) or the carbon source (oil) runs out. The control sample should not show any appreciable microbial growth or physical degradation.

### Conclusions

Students' data should support their conclusions. Students should find that *Penicillium* and *Pseudomonas* break down oil at about the same rate. *Penicillium* and *Pseudomonas* together break up the oil at a slightly faster rate. The control will show no oil breaking up. Students should notice that not all oil will be degraded. That which does not degrade will be incorporated into the food chain.

Because Ms. Childs expressed concerns about the appearance of the beach, students may suggest that the use of *Pseudomonas* is preferable to *Penicillium* because of the appearance of the *Pseudomonas* when the oil is broken down.

**Sample Data Table**

**Density of Microbes**

| Day | Control | Fertilizer | Penicillium | Pseudomonas | Penicillium w/ fertilizer | Pseudomonas w/ fertilizer |
|---|---|---|---|---|---|---|
| 1 | | | | | | |
| 2 | | | Answers will vary. | | | |
| 3 | | | | | | |
| 4 | | | | | | |

HOLT
BIOSOURCES™
LAB PROGRAM

# BIOTECHNOLOGY

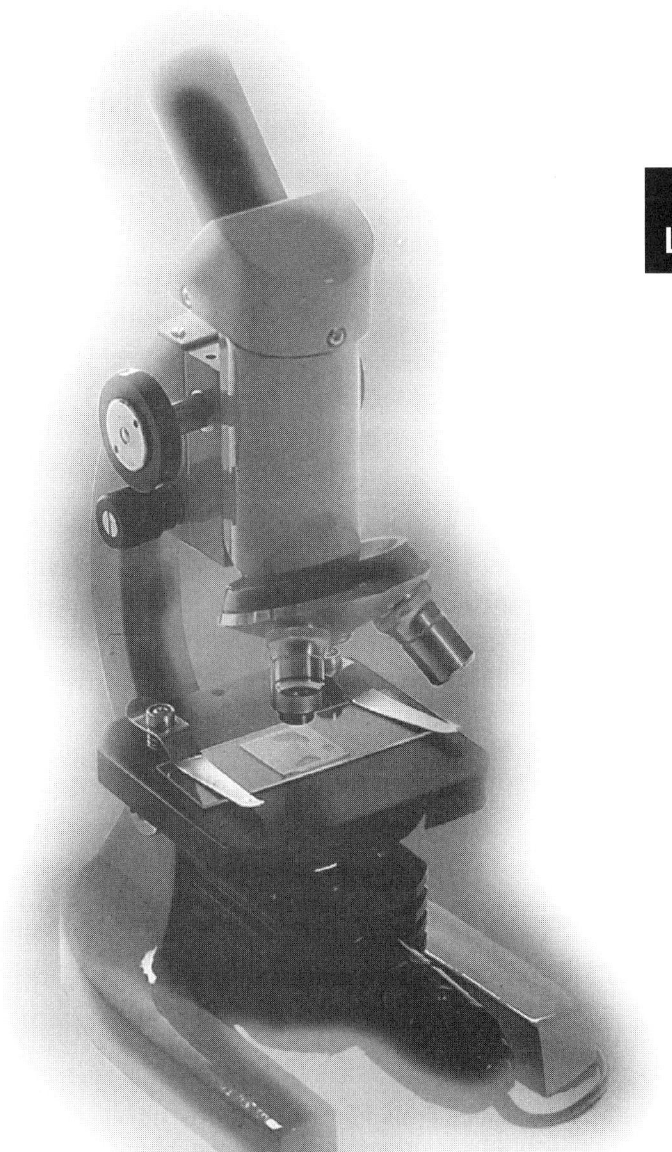

INCLUDES
LABS D1–D12

HOLT, RINEHART AND WINSTON
*Harcourt Brace & Company*
**Austin** • New York • Orlando • Atlanta • San Francisco • Boston • Dallas • Toronto • London

HOLT BIOSOURCES *LAB PROGRAM*

# BIOTECHNOLOGY

## Staff Credits

**Editorial Development**
Carolyn Biegert
Janis Gadsden
Debbie Hix

**Copyediting**
Amy Daniewicz
Denise Haney
Steve Oelenberger

**Prepress**
Rose Degollado

**Manufacturing**
Mike Roche

**Design Development and Page Production**
Morgan-Cain & Associates

## Acknowledgments

**Contributors**
David Jaeger
Will C. Wood High School
Vacaville, CA

George Nassis
Kenneth G. Rainis
WARD'S Natural Science Establishment
Rochester, NY

Suzanne Weisker
Science Teacher and Department Chair
Will C. Wood High School
Vacaville, CA

**Editorial Development**
WordWise, Inc.

**Cover**
Design—Morgan-Cain & Associates
Photography—Sam Dudgeon

**Lab Reviewers**

***Lab Activities***
Ted Parker
Forest Grove, OR

Mark Stallings, Ph.D.
Chair, Science Department
Gilmer High School
Ellijay, GA

George Nassis
Kenneth G. Rainis
Geoffrey Smith
WARD'S Natural Science Establishment
Rochester, NY

***Lab Safety***
Kenneth G. Rainis
WARD'S Natural Science Establishment
Rochester, NY

Jay Young, Ph.D
Chemical Safety Consultant
Silver Spring, MD



Printed in the United States of America

**ISBN 0-03-051407-X**

1 2 3 4 5 6 022 00 99 98 97

# Contents

# Organizing Laboratory Data

Your data are all the records you have gathered from an investigation. The types of data collected depend on the activity. Data may be a series of weights or volumes, a set of color changes, or a list of scientific names. No matter which types of data are collected, all data must be treated carefully to ensure accurate results. Sometimes the data seem to be wrong, but even then, they are important and should be recorded accurately. Remember that nature cannot be wrong, regardless of what you discover in the laboratory. Data that seem to be "wrong" are probably the result of experimental error.

There are many ways to record and organize data, including data tables, charts, diagrams, and graphs. Your teacher will help you decide which format is best suited to the type of data you collect.

It is important to include the appropriate units when you record data. Remember that data are measurements or observations, not merely numbers. Data tables, graphs, and diagrams should have titles that are descriptive and complete enough to ensure that another person could understand them without having been present during the investigation.

Many important scientific discoveries have been made accidentally in the course of an often unrelated laboratory activity. Scientists who keep very careful and complete records sometimes notice unexpected trends in and relationships among data long after the work is completed. The laboratory notebooks of working scientists are studded with diagrams and notes; every step of every procedure is carefully recorded.

## Data Tables and Charts

Data tables are probably the most common means of recording data. Although prepared data tables are often provided in laboratory manuals, it is important that you be able to construct your own. The best way to do this is to choose a title for your data table and then make a list of the types of data to be collected. This list will become the headings for your data columns. For example, if you collected data on plant growth that included both the length of time it took for the plant to grow and the amount of growth, you could record your data in a table like this:

**Plant Growth Data**

| | |
|---|---|
| | |
| | |
| | |
| | |
| | |
| | |

These data are the basis for all your later interpretations and analyses. You can always ask new questions about the data, but you cannot get new data without repeating the experiment.

## Graphs

After data are collected, you must determine how to display them. One way of showing your result is to use a graph. Two types of graphs are commonly used: the line graph and the bar graph. In a line graph, the data are arranged so that two variables are represented as a single point. You could easily make a line graph of the data shown in the growth table. The first step is to draw and label the axes. Before you do this, however, you must decide which column of data should be represented on the *x*-axis (horizontal axis) and which should be represented on the *y*-axis (vertical axis).

Experiments have two types of variables, or factors that may change. Independent variables are variables that could be present even if other factors were not. For the example above, "Time" is an independent variable because time exists regardless of whether plants are present. An independent variable is, by convention, plotted on the *x*-axis of a graph. Dependent variables are variables that change because an independent variable changes. A dependent variable for this example would be "Height of plant." Dependent variables are plotted on the *y*-axis of a graph.

Next you must choose the scale for the axes of your graph. You want the graph to take up as much of the paper as possible because large graphs are much easier to read and make than small ones. For each axis, you must choose a scale that uses the largest amount of graph paper. Remember, once you choose the interval for the scale (the number of days each block represents on the *x*-axis, for example), you cannot change it. You cannot say that block one represents 1 day and block two represents 10 days. If you change the scale, your graph will not accurately represent your data.

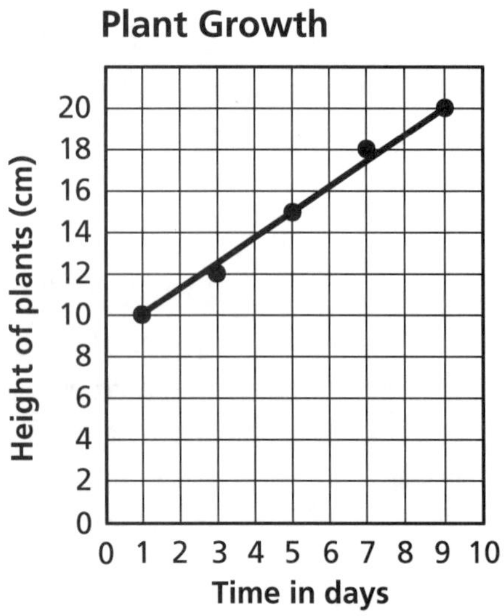


The next step is to mark the points for each pair of numbers. When all points are marked, draw the best straight or curved line between them. Remember that you do not "connect the dots" when you draw a graph. Instead, you should draw a "best fit" curve—a line or smooth curve that intersects or comes as close as possible to your set of data points.

If you choose to represent your data by using a bar graph, the first steps are similar to those for the line graph. You must first choose your axes and label them. The independent variable is plotted on the *x*-axis, and the dependent variable is plotted on the *y*-axis. However, instead of plotting points on the graph, you represent the dependent variable as a bar extending from the *x*-axis to where you would have drawn the points. Using the sample data on plant growth, for example, on day 1, the height of the plant was 10 cm. On your graph, you would make a bar that extends to the height of the 10 cm mark on the *y*-axis.

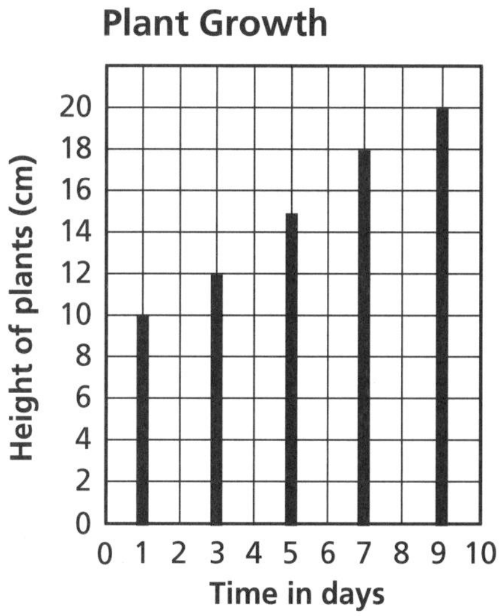


## *Diagrams*

In some cases, the data you must represent are not numerical. That means that they cannot be put into a data table or graphed. The best way to represent this type of information is to draw and label it. To do this, you simply draw what you see and label as many parts or structures as possible. This technique is especially useful in the biology laboratory, where many investigations involve the observation of living or preserved specimens. Remember, you do not have to be an artist to make a good laboratory drawing.

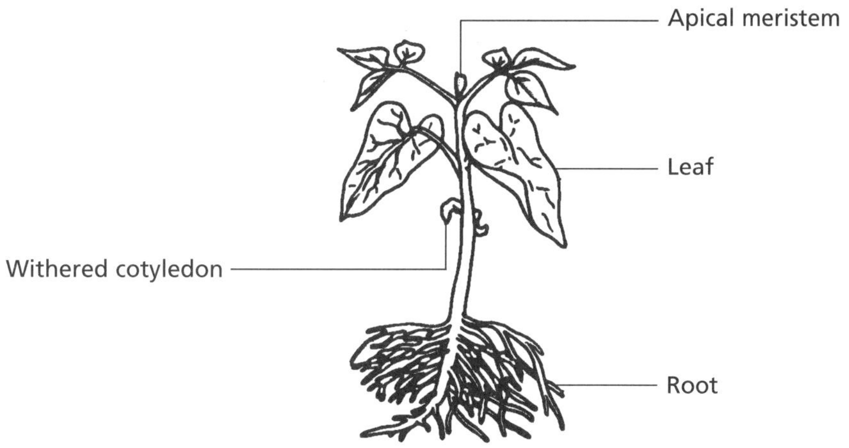


**Bean seedling**

There are several things you need to remember as you make your laboratory drawing. First, make the drawing large enough to be easily studied. Include all of the visible structures in your drawing. Second, drawings should also show the spacing between the parts of the specimen in proportion to its actual appearance. Size relationships are important in understanding and interpreting observations. Third, in order for your drawing to be the most useful to you, you need to label it. All labels should be clearly and neatly printed. Lines drawn from labels to the corresponding parts should be straight, so be sure to use a ruler. Label lines should never cross each other. Finally, be sure to title the drawing. Someone who looks at your drawing should be able to identify the specimens. Remember, neatness and accuracy are the most important parts of any laboratory drawing.

# Laboratory Safety

Your biology laboratory is a unique place where you can learn by doing things that you couldn't do elsewhere. It also involves some dangers that can be controlled if you follow these safety notes and all instructions from your teacher.

It is your responsibility to protect yourself and other students by conducting yourself in a safe manner while in the laboratory. Familiarize yourself with the printed safety symbols—they indicate additional measures that you must take.

## *While in the Laboratory, at All Times . . .*

- **Familiarize yourself with a lab activity—especially safety issues—before entering the lab.** Know the potential hazards of the materials, equipment, and the procedures required for the activity. Ask the teacher to explain any parts you do not understand before you start.
- **Never perform any experiment not specifically assigned by your teacher.** Never work with any unauthorized material.
- **Never work alone in the laboratory.**
- **Know the location of all safety and emergency equipment used in the laboratory.** Examples include eyewash stations, safety blankets, safety shower, fire extinguisher, first-aid kit, and chemical-spill kit.
- **Know the location of the closest telephone,** and be sure there is a posted list of emergency phone numbers, including poison control center, fire department, police, and ambulance.
- **Before beginning work: tie back long hair, roll up loose sleeves, and put on any personal protective equipment as required by your teacher.** Avoid or confine loose clothing that could knock things over, ignite from a flame, or soak up chemical solutions.
- **Report any accident, incident, or hazard—no matter how trivial—to your teacher immediately.** Any incident involving bleeding, burns, fainting, chemical exposure, or ingestion should also be reported to the school nurse or physician.
- **In case of fire, alert the teacher and leave the laboratory.**
- **Never eat, drink, or apply cosmetics.** Never store food in the laboratory. Keep your hands away from your face. Wash your hands at the conclusion of each laboratory activity and before leaving the laboratory. Remember that some hair products are highly flammable, even after application.
- **Keep your work area neat and uncluttered.** Bring only books and other materials that are needed to conduct the experiment.
- **Clean your work area at the conclusion of the lab as your teacher directs.**

- **When called for, use the specific safety procedures below.**

## Eye Safety

- **Wear approved chemical safety goggles as directed.** Goggles should always be worn whenever you are working with a chemical or chemical solution, heating substances, using any mechanical device, or observing a physical process.
- **In case of eye contact**

  (1) Go to an eyewash station and flush eyes (including under the eyelids) with running water for at least 15 minutes.

  (2) Notify your teacher or other adult in charge.
- **Wearing contact lenses for cosmetic reasons is prohibited in the laboratory.** Liquids or gases can be drawn up under the contact lens and into direct contact with the eyeball. If you must wear contact lenses prescribed by a physician, inform your teacher. You must wear approved eye-cup safety goggles—similar to goggles individuals wear when swimming underwater.
- **Never look directly at the sun through any optical device or lens system, or gather direct sunlight to illuminate a microscope.** Such actions will concentrate light rays that will severely burn your retina, possibly causing blindness!

## Electrical Supply

- **Never use equipment with frayed cords.**
- **Ensure that electrical cords are taped to work surfaces** so that no one will trip and fall and so that equipment can't be pulled off the table.
- **Never use electrical equipment around water or with wet hands or clothing.**

## Clothing Protection

- **Wear an apron or lab coat when working in the laboratory to prevent chemicals or chemical solutions from coming in contact with skin or contaminating street clothes.** Confine all loose clothing and long jewelry.

## Animal Care

- **Do not touch or approach any animal in the wild.** Be aware of poisonous or dangerous animals in any area where you will be doing outside fieldwork.
- **Always obtain your teacher's permission before bringing any animal (or pet) into the school building.**
- **Handle any animal only as your teacher directs.** Mishandling or abuse of any animal will not be tolerated!

## Sharp Object Safety

- **Use extreme care with all sharp instruments, such as scalpels, sharp probes, and knives.**
- **Never use double-edged razor blades in the laboratory.**
- **Never cut objects while holding them in your hand.** Place objects on a suitable work surface.

## *Chemical Safety*

- **Always wear appropriate personal protective equipment.** Safety goggles, gloves, and an apron or lab coat should always be worn when working with any chemical or chemical solution.
- **Never taste, touch, or smell any substance or bring it close to your eyes, unless specifically told to do so by your teacher.** If you are directed by your teacher to note the odor of a substance, do so by waving the fumes toward you with your hand. Never pipet any substance by mouth; use a suction bulb as directed by your teacher.
- **Always handle any chemical or chemical solution with care.** Check the label on the bottle and observe safe-use procedures. Never return unused chemicals or solutions to their containers. Return unused reagent bottles or containers to your teacher. Store chemicals according to your teacher's directions.
- **Never mix chemicals** unless specifically told to do so by your teacher.
- **Never pour water into a strong acid or base.** The mixture can produce heat and can splatter. Remember this rhyme:

  "Do as you oughta—
  Add acid (or base) to water."
- **Report any spill immediately to your teacher.** Handle spills only as your teacher directs.
- **Check for the presence of any source of flames, sparks, or heat (open flame, electric heating coils, etc.) before working with flammable liquids or gases.**

## *Plant Safety*

- **Do not ingest any plant part used in the laboratory (especially seeds sold commercially).** Do not rub any sap or plant juice on your eyes, skin, or mucous membranes.
- **Wear protective gloves (disposable polyethylene gloves) when handling any wild plant.**
- **Wash hands thoroughly after handling any plant or plant part (particularly seeds).** Avoid touching your hands to your face and eyes.
- **Do not inhale or expose yourself to the smoke of any burning plant.**
- **Do not pick wildflowers or other plants unless directed to do so by your teacher.**

## *Proper Waste Disposal*

- **Clean and decontaminate all work surfaces and personal protective equipment as directed by your teacher.**
- **Dispose of all sharps (broken glass and other contaminated sharp objects) and other contaminated materials (biological and chemical) in special containers as directed by your teacher.**

## *Hygienic Care*

- **Keep your hands away from your face and mouth.**
- **Wash your hands thoroughly before leaving the laboratory.**
- **Remove contaminated clothing immediately; launder contaminated clothing separately.**
- **When handling bacteria or similar microorganisms, use the proper technique demonstrated by your teacher.** Examine microorganism cultures (such as petri dishes) without opening them.
- **Return all stock and experimental cultures to your teacher for proper disposal.**

## *Heating Safety*

- **When heating chemicals or reagents in a test tube, never point the test tube toward anyone.**
- **Use hot plates, not open flames.** Be sure hot plates have an "On-Off" switch and indicator light. Never leave hot plates unattended, even for a minute. Never use alcohol lamps.
- **Know the location of laboratory fire extinguishers and fire blankets.** Have ice readily available in case of burns or scalds.
- **Use tongs or appropriate insulated holders when heating objects.** Heated objects often do not look hot. Never pick up an object with your hands unless you are certain it is cold.
- **Keep combustibles away from heat and other ignition sources.**

## *Hand Safety*

- **Never cut objects while holding them in your hand.**
- **Wear protective gloves when working with stains, chemicals, chemical solutions, or wild (unknown) plants.**

## *Glassware Safety*

- **Inspect glassware before use; never use chipped or cracked glassware.** Use borosilicate glass for heating.
- **Do not attempt to insert glass tubing into a rubber stopper without specific instruction from your teacher.**
- **Always clean up broken glass by using tongs and a brush and dustpan.** Discard the pieces in an appropriately labeled "sharps" container.

## *Safety With Gases*

- **Never directly inhale any gas or vapor.** Do not put your nose close to any substance having an odor.
- **Handle materials prone to emit vapors or gases in a well-ventilated area.** This work should be done in an approved chemical fume hood.

# Using Laboratory Techniques and Experimental Design Labs

You will find two types of laboratory exercises in this book.

1. ***Laboratory Techniques*** labs help you gain skill in biological laboratory techniques.
2. ***Experimental Design*** labs require you to use the techniques learned in *Laboratory Techniques* labs to solve problems.

## Working in the World of a Biologist

Laboratory Techniques and Experimental Design labs are designed to show you how biology fits into the world outside of the classroom. For both types of labs, you will play the role of a working biologist. You will gain experience with techniques used in biological laboratories and practice real-world work skills, such as creating a plan with available resources, working as part of a team, developing and following a budget, and writing business letters.

### Tips for success in the lab

Preparation helps you work safely and efficiently. Whether you are doing a Laboratory Techniques lab or an Experimental Design lab, you can do the following to help ensure success.

- **Read a lab twice** before coming to class so you will understand what to do.
- **Read and follow the safety information** in the lab and on pages viii–xi.
- **Prepare data tables** before you come to class.
- **Record all data and observations immediately** in your data tables.
- **Use appropriate units** whenever you record data.
- **Keep your lab table organized** and free of clutter.

## Laboratory Techniques Labs

Each Laboratory Techniques lab enables you to practice techniques that are used in biological research by providing a step-by-step procedure for you to follow. You will use many of these techniques later in an Experimental Design lab. The parts of a Laboratory Techniques lab are described below.

1. *Skills* identifies the techniques and skills you will learn.
2. *Objectives* tells you what you are expected to accomplish.
3. *Materials* lists the items you will need to do the lab.
4. *Purpose* is the setting for the lab.
5. *Background* is the information you will need for the lab.
6. *Procedure* provides step-by-step instructions for completing the lab and reminders of the safety procedures you should follow.
7. *Analysis* items help you analyze the lab's techniques and your data.
8. *Conclusions* items require you to form opinions based on your observations and the data you collected in the lab.

9. *Extensions* items provide opportunities to find out more about topics related to the subject of the lab.

## Experimental Design Labs

Each of these labs requires you to develop your own procedure to solve a problem that has been presented to your company by a client. The procedures you develop will be based on the procedures and techniques you learned in previous Laboratory Techniques labs. You must also decide what equipment to use for a project and determine the amount you should charge the client. The parts of an Experimental Design lab are described below.

1. *Prerequisites* tells you which Laboratory Techniques labs contain procedures that apply to the Experimental Design lab.
2. *Review* tells you which concepts you need to understand to complete the lab.
3. The *Letter* contains a request from a client to solve a problem or to do a project.
4. The *Memorandum* is a note from a supervisor directing you to perform the work requested by the client, and providing clues or directions that will help you design a successful experiment or project. The *Memorandum* also contains the following: a *Proposal Checklist,* which must be completed before you start the lab; *Report Procedures,* which tells you what should be included in your lab report; *Required Precautions,* which indicate the safety procedures you should follow during the lab; and *Disposal Methods,* which tells you how to dispose of the materials used.
5. *Materials and Costs* is a list of what you might need to complete the work and the unit cost of each service and item.

### What you should do before an Experimental Design lab

Before you will be allowed to begin an Experimental Design lab, you must turn in a proposal that includes the question to be answered, the procedure you will use, a detailed data table, and a list of all the proposed materials and their costs. Before you begin writing your proposal, follow these steps.

- **Read the lab thoroughly,** and jot down any clues you find that will help you successfully complete the lab.
- **Consider what you must measure or observe** to solve the problem.
- **Think about *Laboratory Techniques* labs you have done that required similar** measurements and observations.
- **Imagine working through a procedure,** keeping track of each step and of the equipment you will need.

### What you should do after an Experimental Design lab

After you finish, prepare a report as described in the *Memorandum.* The report can be in the form of a one- or two-page letter to the client, plus an invoice showing the cost of each phase of the work and the total amount you charged the client. Carefully consider how to convey the information the client needs to know. In some cases, graphs and diagrams may communicate information better than words can.

Name ______________________________

Date ______________ Class ______________

HOLT **BIOSOURCES**
LAB PROGRAM
BIOTECHNOLOGY

# D1 Laboratory Techniques: Staining DNA and RNA

## Skills

PREPARATION NOTES

- making a squash of onion root tips *(Allium sativum)*
- staining the nucleic acids of a typical plant cell
- using a compound light microscope

## Objectives

**Time Required:** one 50-minute period

- *Demonstrate* how to make a squash of onion root tips.
- *Compare* the location in the cell of DNA and RNA.

## Materials

**Materials**
Materials for this lab activity can be purchased from WARD'S. See the *Master Materials List* for ordering instructions.

- safety goggles
- lab apron
- microscope slide
- methyl green-pyronin Y stain in dropper bottle
- forceps
- vial of pretreated *Allium* root tips
- wooden macerating stick
- watch or clock
- coverslip
- paper towels
- pencil with eraser
- compound light microscope
- prepared reference slide of plant cells
- prepared reference slide of animal cells
- mounting medium (Piccolyte II) in dropper bottle (optional)

## Purpose

You have just come back from a visit to a pathology laboratory, where you observed a renal (kidney) biopsy. The pathologist wanted to determine if the person had rejected a kidney she recently received as a transplant. A methyl green-pyronin Y stain was used to see if lymphocyte-type cells were collecting around the blood vessels—a sign of possible organ rejection. These cells have RNA in their cytoplasm and when stained become bright pink. The pathologist has given you some stain to try on plant cells and mentioned that you can use the stain for detecting both DNA and RNA. You are going to see if the pathologist is correct and if it is difficult to see the difference in staining results.

## Background

**Additional Background**
Nucleic acids, along with proteins, fats, and carbohydrates, are the four major groups of organic chemicals that make up cells in an organism. Nucleic acids are responsible for storing information about the structure of proteins and the genetic makeup of an organism. Nucleic acids also control the reproduction of cells.

Root tip cells of onions (*Allium sativum*) are frequently used to study DNA and RNA in plant cells. In plants, mitosis occurs in special growth regions called **meristems** located at the tips of the roots and stems. To observe chromosomes in stem and root meristems, biologists prepare a special kind of slide called a **squash.** This preparation is just what it sounds like. Tissue containing actively dividing cells is removed from a root or stem meristem and treated with hydrochloric acid to fix the cells, or to stop them from dividing. The cells are then stained, made into a wet mount, and squashed and spread into a single layer by applying pressure to the coverslip.

The stain methyl green-pyronin Y is a mixture of two different stains. Methyl green is absorbed by DNA only and stains the DNA blue. Pyronin Y is absorbed by RNA only and stains the RNA pink. Therefore, methyl green-pyronin Y can be used to differentiate between the two nucleic acids.

## Procedure

A nucleic acid contains only five elements: carbon, hydrogen, oxygen, nitrogen, and phosphorus. These elements combine in three distinct subunits to form a nucleotide, the smallest piece of a nucleic acid strand. Nucleotides contain one of two sugars. Ribose is found only in ribonucleic acid (RNA). Deoxyribose is found only in deoxyribose nucleic acids (DNA).

### Preparation Tip

Purchase the WARD'S kit, which includes onion root tips and ready-made 0.5% methyl green and 0.1% pyronin Y in acetate buffer.

### Disposal

Methyl green and pyronin Y are water-based stains. They can be washed down the drain with water.

### PROCEDURAL NOTES

### Safety Precautions

- Have students wear safety goggles and lab aprons.
- Discuss all safety symbols and caution statements with students.
- Review the rules for carrying and using the compound microscope.
- Remind students to notify you immediately of any chemical spills. Also caution them to never taste, touch, or smell any substance or bring it close to their eyes.

1. Put on safety goggles and a lab apron.

2. Place a microscope slide on a paper towel on a smooth, flat surface. **CAUTION: Glassware is fragile. Notify your teacher promptly of any broken glass or cuts. Do not clean up broken glass or spills unless your teacher tells you to do so.** Add two drops of methyl green-pyronin Y stain to the center of the slide. **CAUTION: Methyl green-pyronin Y stain will stain your skin and clothing. Promptly wash off spills to minimize staining.**

3. Use forceps to transfer a prepared onion root tip to the drop of stain on the microscope slide.

4. Carefully smash the root tip by gently but firmly tapping the root with the end of a wooden macerating stick. *Note: Tap the macerating stick in a straight up-and-down motion.*

5. Allow the root tip to stain for 10 to 15 minutes. *Note: Do not let the stain dry. Add more stain if necessary.*

6. Place a coverslip over your preparation, and cover the slide by folding a paper towel over it. Using the eraser end of a pencil, gently, but forcefully, press straight down (with no twisting) on the coverslip through the paper towel. Apply only enough pressure to squash the root tip into a single cell layer. *Note: Be very careful not to move the coverslip while you are pressing down with the pencil eraser. Also be very careful not to press too hard. If you press too hard, you might break the glass slide and tear apart the cells in the onion root tip.*

   ♦ Why do you squash and spread out the root tip?

   to make the tissue one cell layer thick so that the contents of individual cells can be seen under a microscope

7. Examine your prepared slide under both the low power and the high power of a compound light microscope. *Note: Remember that your mount is fairly thick, so be careful not to switch to the high-power objective too quickly. You may shatter the coverslip and destroy your preparation. You will need to focus up and down carefully with the fine adjustment to better see the structures under study.* Complete the data table on the next page.

**Procedural Tips**

- Students should be able to locate DNA (stained blue-purple) and RNA (stained pink-red), as well as a nucleus, multiple nucleoli, and granular cytoplasmic bodies, in the onion cells. Students may also be able to locate granular chromatin and condensed chromosomes in cells in the process of division.
- The onion root tip preparation the students make may be fairly thick. Caution students to take care not to force the objective down onto the slide they prepare, either crushing the slide or damaging the high-power objective.
- Place prepared reference slides of DNA/RNA in animal cells and nucleic acids in plant cells, in focus, under low power on a class-accessible microscope. If possible, project the slides on a projecting scope.

### *Color of Cell Structures in Onion Root Tip Cells*

| Structure | Stained color |
|---|---|
| nucleus | dark reddish-purple |
| nucleolus | red, pink |
| cytoplasm | light pink |
| chromosomes | blue, purple |

**8.** In the space below, draw and label a representative plant cell from your prepared slide. Include all visible organelles. Indicate where DNA and RNA are found in the cell.

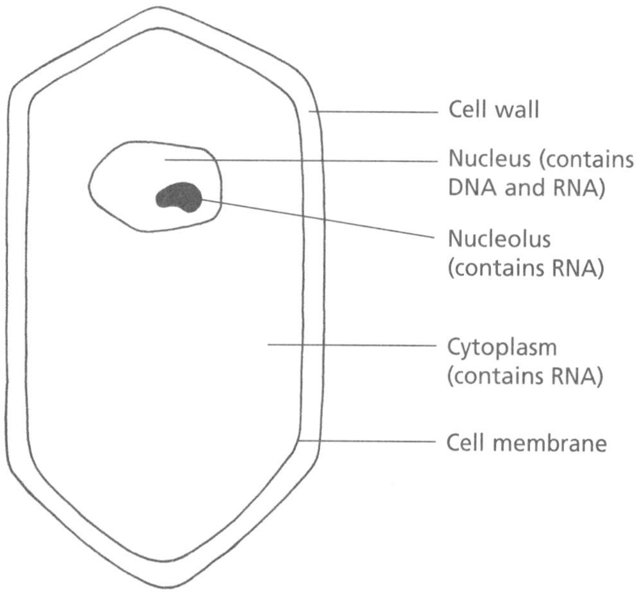


**9.** Observe the prepared reference slides of plant and animal cells. Compare them with the slide you prepared.

♦ How does your slide compare with the prepared slide of onion root tip cells?

Answers will vary depending on students' technique.

**10.** Dispose of your materials according to the directions from your teacher.

**11.** Clean up your work area and wash your hands before leaving the lab.

## Analysis

**12.** What color did the deoxyribonucleic acid (DNA) stain in your root tip squash?

Methyl green-pyronin Y stains DNA blue to purple.

**13.** How do you know this material is DNA?

It is located in the nucleus, but not within the nucleolus, where RNA is found.

**14.** What color did the nucleoli appear in the stained slide? What does this tell you about the composition of nucleoli?

Methyl green-pyronin Y stains nucleoli red to pink. Nucleoli contain RNA.

**15.** Did each nucleus have only one nucleolus or several? What appeared to be the most common number of nucleoli?

Each nucleus will have at least one nucleolus, but most will have two or three.

**16.** Were you able to see any cells in the process of mitotic division? If so, what did the cells look like?

Students should be able to locate several cells in the process of mitotic division. The nuclear material will appear to have condensed into short dark chromosomes stained blue.

## Conclusions

**17.** Where is DNA located in both plant and animal cells?

DNA is found in the same location in both plant and animal cells; DNA is found only in the chromosomes of the nucleus.

**18.** Where is RNA located in both plant and animal cells?

RNA is found in the same locations in both plant and animal cells; RNA is found in the nucleus, nucleoli, and cytoplasm.

## Extensions

**19.** Make a permanent slide of your root tip squash preparation. Ask your teacher to provide you with a mounting medium (Piccolyte II). To make a permanent slide, remove the coverslip from your wet mount. Add a drop of the mounting medium, then replace the coverslip. Place the slide on a flat surface and allow it to dry for several days.

**20.** Use library references to research other staining techniques.

Name ______________________________

Date ________________ Class ________________

HOLT BIOSOURCES LAB PROGRAM BIOTECHNOLOGY

# D2 Laboratory Techniques: Extracting and Spooling DNA

## Skills

- extracting DNA from an animal cell
- spooling DNA

**PREPARATION NOTES**

## Objectives

**Time Required:** one 50-minute period

- *Separate* and *collect* the DNA from bovine liver cells.
- *Describe* the appearance of DNA extracted from a cell.
- *Relate* the location of DNA in a cell to procedures for extracting it.

## Materials

**Materials**

Materials for this lab activity can be purchased from WARD'S. See the *Master Materials List* for ordering instructions.

- safety goggles
- lab apron
- bovine liver (2 cm square)
- mortar and pestle
- fine sand
- graduated cylinder
- SDS/NaCl solution (10 mL)
- cheesecloth (several pieces, 12 cm × 12 cm)
- funnel
- test tube
- ice-water bath
- test-tube rack
- 70% ethanol (4 mL)
- inoculating loop

## Purpose

**Preparation Tips**

- Beef liver can be purchased from a local grocery store or butcher shop. Cut the liver into 2 cm cubes and freeze until ready to use.
- Prepare or purchase 100 mL of 10% SDS/1.5% NaCl solution.

**Disposal**

- Dilute solutions in a ratio of 1 part solution to 20 parts water, and flush the diluted solutions down the drain with water.
- Wrap leftover beef liver and spooled DNA in old newspaper, and place in the trash.

You are an intern working in the city's forensics lab. You will be assisting the forensics technician with many of her routine laboratory tests and procedures. One procedure the technician does frequently is extract DNA from cells and purify it. The purified DNA is used to help prepare a DNA fingerprint to help solve crimes. To make sure you know how to do this procedure correctly, the technician has asked you to extract DNA from the cells of a piece of bovine (beef) liver and spool the DNA for observation.

## Background

The extraction of DNA from cells and its purification are of primary importance to the field of biotechnology. Extraction and purification of DNA are the first steps in the analysis and manipulation of DNA that allow scientists to detect genetic disorders, produce DNA fingerprints of individuals, and even create genetically-engineered organisms used to produce beneficial products such as insulin, antibiotics, and hormones.

The process of extracting DNA, regardless of its original source, involves the following steps. The first step in extracting DNA from a cell is to **lyse,** or break open, the cell. One common way to lyse cells is to grind a piece of tissue along with a mild abrasive material in a mortar with a pestle. After the cells have been broken open, a solution containing salt (NaCl) and a detergent containing the compound SDS, or sodiumdodecyl sulfate, is used to break down and emulsify the fat and proteins that make up the cell membrane. Finally, ethanol is added. Because DNA is soluble in water, the addition of ethanol causes the DNA to **precipitate,** or settle out of solution, leaving behind all remaining cellular components that are not soluble in ethanol. Finally, the DNA can be spooled, or wound onto an inoculating loop, and pulled from the test tube.

## Procedure

### PROCEDURAL NOTES

**Safety Precautions**

- Have students wear safety goggles and lab aprons.
- Discuss all safety symbols and caution statements with students.
- Review the rules for carrying and using the compound microscope.
- Caution students to never use ethanol near an open flame.

**Procedural Tip**

Tell students that the DNA they are spooling is not pure DNA. It contains various cellular debris. Purifying the DNA is a long and detailed process.

1. Put on safety goggles and a lab apron.
2. Place a piece of bovine liver in a mortar. Add several grains of sand.
3. Pour 10 mL SDS/NaCl solution in the mortar.
4. Use a pestle to grind the ingredients until they form a thick fluid. *Note: Be careful not to overgrind this mixture.*
5. Place several layers of cheesecloth into a funnel. Pour the contents of the mortar through the cheesecloth into a test tube until it contains at least 2 mL of the extract. *Note: You may need to gently squeeze the cheesecloth to remove all the fluid from the cheesecloth.* **CAUTION: Glassware is fragile. Notify your teacher promptly of any broken glass or cuts. Do not clean up broken glass or spills unless your teacher tells you to do so.**
6. Place the test tube in an ice-water bath.
7. Measure 4 mL of *ice-cold* ethanol in a clean graduated cylinder.
8. Hold the test tube at a 45° angle. Slowly pour the 4 mL of ice-cold ethanol into the tube. Be careful to pour the ethanol slowly down the side of the tube. *Note: Do not pour the ethanol too fast or directly into the liver solution.* As you pour the ethanol into the test tube, observe the interface line, as shown in the diagram below.

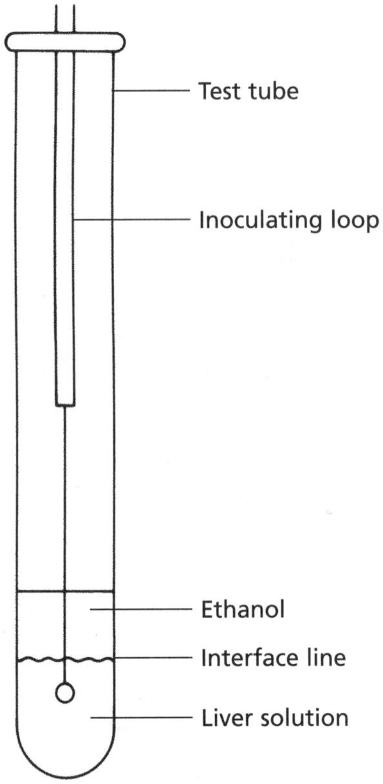

9. Gently insert an inoculating loop into the test tube as far as the interface line. Carefully and slowly move the loop in circles, as shown in the diagram at right. This motion spools the long threads of DNA around the end of the loop. *Note: Spool just enough DNA so that you can see it and observe its physical characteristics.* Lift the inoculating loop out of the solution in the test tube. While the DNA is being pulled out of the test tube, try stretching it. Then dip the inoculating loop again to get more DNA.

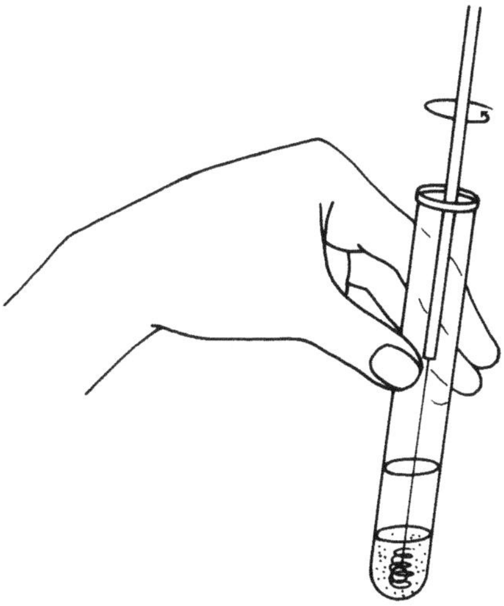

10. When spooling is complete, pull the inoculating loop from the test tube, and return the test tube to a test-tube rack.

11. Dispose of your materials according to the directions from your teacher.

12. Clean up your work area and wash your hands before leaving the lab.

## Analysis

13. Describe the appearance of the DNA you spooled from the crushed bovine liver cells.

Answers will vary. Students should describe the color (clear or white) and the viscosity (like mucus). The length of the DNA may differ; it may be several short strands or a single giant thread. Some students may touch the DNA and describe its sticky texture. Some may also mention that the strands are elastic.

14. What was the purpose of adding sand to the liver in the mortar?

The sand serves as a mild abrasive. Grinding the liver along with the sand using a pestle breaks open, or lyses, the cell, the first step in the extraction of DNA.

15. What happens to the cell when the SDS/NaCl solution is added to the liver mixture in the mortar?

The addition of SDS/NaCl solution breaks down and emulsifies the fat and proteins that make up the cell membrane, making it easier to precipitate the DNA out of solution.

**16.** What was the purpose of filtering the liver mixture through cheesecloth?

Filtration with the cheesecloth physically separates the DNA from the cell membrane, protein, and sand used to grind the liver.

**17.** What was done to the DNA so that it could be observed and spooled?

Ethanol is added to the filtered liver mixture to precipitate the DNA out of solution before it was spooled onto a loop.

**18.** How can you determine whether the material pulled from the test tube was DNA?

Stain the material with a stain that indicates DNA.

## Conclusions

**19.** How is DNA protected inside an animal cell? How does this location relate to the procedure you used in this lab to extract DNA?

The DNA in animal cells is protected by the nuclear membrane and the cell membrane. The procedures of grinding the cells and breaking down the membranes enabled the DNA to be released from the cell nucleus.

**20.** Biotechnologists research DNA. How could the procedure you used today facilitate their research?

This procedure could be used to extract DNA from a cell so that it can be studied further or prepared for work with recombinant DNA.

## Extensions

**21.** Find out what a DNA fingerprint is and how it is used to compare samples of DNA from different sources. Explain how the technique you learned in today's lab is used in developing a DNA fingerprint.

**22.** Find out about the Human Genome Project. What is this project attempting to do? How would the procedure used in this lab be used as part of the Human Genome Project?

Name ______________________________

Date ______________ Class ______________

HOLT BIOSOURCES
LAB PROGRAM
BIOTECHNOLOGY

# D3 Laboratory Techniques: Genetic Transformation of Bacteria

## Skills

- using aseptic technique
- predicting experimental results
- growing bacteria on agar

**PREPARATION NOTES**

## Objectives

**Time Required:** 2 or 3 class periods

- *Introduce* a plasmid into bacterial cells to genetically transform the bacteria.
- *Evaluate* whether bacterial cells were transformed by observing the characteristics of bacteria grown on petri dishes that contain agar with X-gal.

## Materials

**Materials**

Materials for this lab activity can be purchased from WARD'S. See the *Master Materials List* for ordering instructions.

**Additional Materials for Teacher Prep**

- insulated gloves
- thermometer
- graduated cylinder
- Bunsen burner and striker
- 5 tryptic soy agar slants
- 6 sterile cotton swabs
- autoclave

- safety goggles
- gloves
- lab apron
- disinfectant solution in squeeze bottle
- paper towels
- nontoxic permanent marker
- 15 mL sterile plastic tubes with lids (2)
- ice bath
- 1 mL sterile plastic serological pipets with 0.01 mL graduations (6)
- pipet bulb
- 0.5 mL of cold 0.5 M $CaCl_2$
- disposable inoculating loops (2)
- stock culture of *E. coli* JM101
- sterile graduated pipets (4)
- 0.01 mL of plasmid pUC8 in a 1.5 mL microtube
- petri dishes with Luria broth agar (2)
- petri dishes with Luria broth + X-gal agar (2)
- plastic foam tube holder
- 42°C water bath
- test-tube rack
- 5 mL of Luria broth
- cotton swabs (4)
- transparent tape or parafilm
- incubator
- biohazard waste container

## Purpose

You are a geneticist who is involved in researching the cause of lactose intolerance, which is a digestive deficiency in humans characterized by the inability to digest lactose, a milk sugar. You decide to conduct an experiment using bacterial cells that cannot metabolize lactose. You design an experiment to genetically change the bacteria so they can digest lactose. Today you will test your experimental design to determine if it works.

## Background

**Additional Background**

The process of genetic transformation was first discovered in 1928 when researchers injected a strain of mice with a mixture of a nonvirulent strain of *Streptococcus pneumoniae* and a heat-killed virulent strain.

***Escherichia coli,*** or ***E. coli,*** is a common bacterium found in the intestines of many mammals. *E. coli* can be classified as lac+ or lac− depending upon its ability to digest lactose. The enzyme **β-galactosidase** breaks down lactose into glucose and galactose for energy consumption. If *E. coli* produces β-galactosidase, it can metabolize lactose and is **lac+**. If *E. coli* is not capable of producing β-galactosidase, it cannot metabolize lactose and is **lac−.** Cells of the JM101 strain of *E. coli* are not capable of producing β-galactosidase and are lac−.

The mice were killed. Neither of these injections given separately killed the mice. Although the researchers did not know it at the time, an exchange of genetic material had occurred between the dead cells and the live cells.

Plasmids are vectors that carry genetic information into bacteria. Restriction enzymes split plasmid DNA at a cleavage site. A fragment of DNA from another source can be joined to the plasmid. The plasmid can be mixed with specially prepared bacterial cells. The bacterial cells will take in the genetic material contained in the plasmid and be transformed. Plasmid pUC8 has the gene for β-galactosidase and an ampicillin resistance gene.

Bacteria normally have two types of DNA, a main chromosome and a circular DNA molecule called a **plasmid.** Plasmids contain only a few genes and are used in genetic engineering to insert genes into other organisms. Genetic engineers refer to plasmids by code numbers. The plasmid pUC8 contains genetic instructions for synthesizing β-galactosidase.

**Genetic transformation** is the process of changing an organism by transferring genetic material from another organism. The plasmid pUC8 contains DNA that will transform *E. coli* JM101 cells into cells that can metabolize lactose. Bacterial cells are more permeable to DNA (more likely to allow DNA to pass through their cell walls and membranes) if they are treated with a calcium chloride solution and exposed to low and high temperatures. The sudden temperature change creates a flow into the bacterial cell, bringing the plasmid into the bacterial cell. The DNA from the plasmid flows through the bacterial cell membrane and wall more easily after treatment.

The compound **X-galactoside,** or **X-gal,** is metabolized by β-galactosidase in a manner similar to the metabolism of lactose. A waste product of X-gal metabolism is a bright blue color. If β-galactosidase is present in bacterial cells, X-gal is metabolized and a bright blue ring forms around the colonies. *E. coli* has been transformed from lac− to lac+ if the colonies turn bright blue after growing on a nutrient medium treated with X-gal.

## Procedure

### Preparation Tips

- Always use aseptic, or sterile, techniques when preparing stock solutions.
- The WARD'S Phenotype Expression kit contains all of the materials necessary for this activity. To prepare the solutions, media, and cultures for this activity, follow the instructions enclosed with the kit.
- Label the petri dishes with Luria broth agar "LB." Label the petri dishes with Luria broth + X-gal agar "LB + X-gal."
- To prepare a dilute household bleach disinfectant solution, add 100 mL household bleach to 900 mL water. Pour into squirt bottles labeled "disinfectant solution."
- Provide separate containers for disposal of micropipetter tips, inoculating loops, pipets, and petri dishes.

### *Part 1—Transformation Technique*

1. Predict the results of a successful transformation by placing a check mark (✔) in the appropriate spaces in the table below. In the table, the presence of or absence of the plasmid pUC8 is shown with a "+" or a "−," respectively.

| Dish | Conditions under which *E. coli* JM101 is grown | Evidence of bacterial growth | Evidence of X-gal digestion |
|---|---|---|---|
| Dish 1 | +pUC8 | ✔ | |
| Dish 2 | +pUC8 and X-gal | ✔ | ✔ |
| Dish 3 | −pUC8 | ✔ | |
| Dish 4 | −pUC8 and X-gal | ✔ | |

2. Put on safety goggles, gloves, and a lab apron.
3. Use aseptic technique throughout this lab. Clean the lab-table surface with disinfectant solution and paper towels.
4. Using a permanent marker, label the lids of two 15 mL plastic tubes with the initials of everyone in your lab group. Then write +pUC8 on one lid and −pUC8 on the other lid. Place the unopened tubes in an ice bath. The tubes, bacteria, and plasmid must always be kept on ice unless the instructions state otherwise.

- Set up 42°C hot-water baths for students ahead of time.

**Disposal**

- To clean bacterial spills, wear gloves. Cover the area with a layer of paper towels. Wet the paper towels with undiluted bleach; allow to stand for 15–20 minutes. Wearing gloves and using forceps, place the residue in a biohazard bag. If broken glass is present, use a brush and dustpan to collect the material, and place it in a suitably marked container.
- When students have completed the lab, all materials that come in contact with bacteria should be sterilized in an autoclave or pressure cooker at 121°C for 15 minutes.
- Package all sharp instruments in a separate metal container for disposal.
- Place disposal bags, following autoclaving, in an outer-sealed, container or bucket with a lid.
- Absorb free liquids and gels with paper towels to minimize risk of leakage.
- Contaminated materials that are to be decontaminated away from the laboratory must be placed in a durable, leakproof container that is closed prior to removal from the laboratory.
- All surfaces must be cleaned and decontaminated with the disinfectant at the conclusion of each lab.

**5.** Using a 1 mL sterile plastic serological pipet, transfer 0.25 mL of cold 0.5 M $CaCl_2$ into each 15 mL plastic tube. **CAUTION: If you get a chemical on your clothing, wash it off at the sink while calling to your teacher.** Place the lids on the tubes securely, and immediately place them in the ice bath.

**6.** Take one plastic tube from the ice bath, and use a disposable inoculating loop to transfer several colonies from the stock culture of *E. coli* into the tube. Vigorously tap the inoculating loop against the wall of the tube to dislodge the cell mass. Then use a sterile plastic graduated pipet to gently mix the bacteria with the $CaCl_2$ solution, making sure the bacteria are suspended and that no cell mass is left on the side of the tube. *Note: Be careful not to transfer any agar. Impurities in agar can inhibit transformation.* Using a new inoculating loop and sterile pipet, repeat this process with the second tube. Check to make sure that the suspension is homogenous and that the lid on each tube is secure. Return each tube to the ice bath when the transfer is complete, and incubate the tubes on ice for 30 minutes.

**7.** Using a 1 mL sterile plastic serological pipet, add 0.01 mL of pUC8 to the +pUC8 tube *only. Very gently* tap the tube with your finger to mix the plasmid into the cell suspension, and return the tube to the ice bath. Leave both tubes in the ice bath for 20 minutes.

**8.** While the tubes are in the ice bath, obtain two petri dishes labeled *LB* (with Luria broth agar) and two petri dishes labeled *LB + X-gal* (with Luria broth + X-gal agar). Label the bottoms of the four petri dishes, as seen in the drawing at right, with the initials of each group member and the information below:

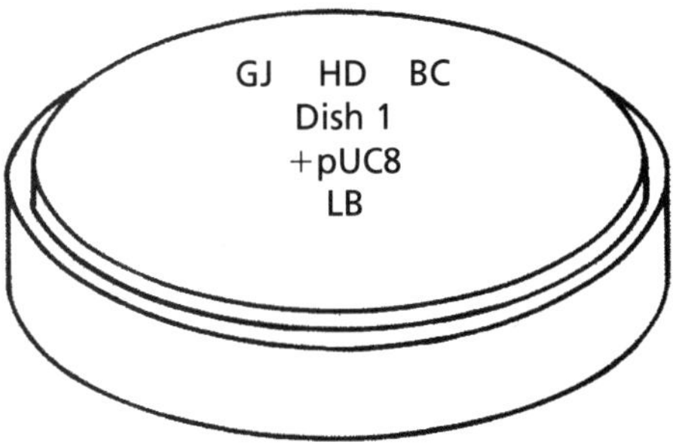


| Dishes labeled *LB:* | Dishes labeled *LB + X-gal:* |
|---|---|
| Dish 1 +pUC8 | Dish 2 +pUC8 |
| Dish 3 –pUC8 | Dish 4 –pUC8 |

**9.** After 20 minutes, remove the tubes from the ice bath and place each tube in a hole in a piece of plastic foam. Heat shock the tubes by floating them in a 42°C water bath for 60 seconds.

**10.** Remove the tubes after 90 seconds, and immediately place them back in the ice bath for 2 minutes. After 2 minutes, remove the tubes from the ice bath, and place them in a test-tube rack to return to room temperature.

**11.** Using a sterile plastic graduated pipet, aseptically transfer 2.5 mL of Luria broth into one tube. Gently tap the tube to mix the contents. Repeat for the second tube. Incubate the tubes at 37°C for 30 minutes.

**PROCEDURAL NOTES**

**Safety Precautions**

- Discuss all safety symbols and caution statements with students.
- Have students wear a lab apron, gloves, and safety goggles.
- Discuss the importance of using aseptic technique, sterilizing lab tables before and after the lab, and placing all contaminated materials in a biohazard bag.

**Procedural Tips**

- Review aseptic technique and general bacteriology techniques.
- Remind students to use a pipet bulb when pipetting. Review or model the proper techniques for using pipet bulbs and serological pipets.
- If class time runs out before the inoculated agar has absorbed the liquid, you may wish to wait to invert the petri dishes until after class.
- The number of colonies will vary by group. Students should gain practice by counting individual colonies. The colonies should be off-white in color and have a round shape.

**12.** Using a 1 mL sterile plastic serological pipet, place 0.25 mL from the tube labeled +pUC8 onto Dish 1. Use a new serological pipet to place 0.25 mL from the tube labeled +pUC8 onto Dish 2. Immediately spread the solution evenly over the agar in Dish 1 using a sterile cotton swab, as seen in the diagram at right. Repeat the procedure for Dish 2. Dispose of the cotton swabs and pipets according to the directions from your teacher.

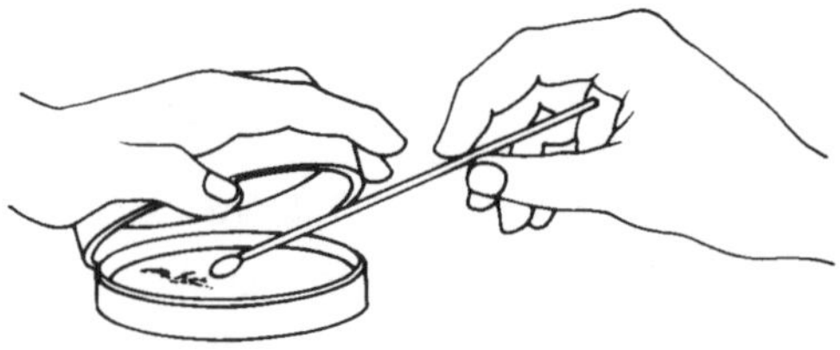

**13.** Repeat step 12 using the −pUC8 tube, Dish 3, and Dish 4.

**14.** Allow the petri dishes to set for 10 minutes or until the liquid has been completely absorbed by the agar.

**15.** Seal each petri dish using two pieces of transparent tape or parafilm. Then invert the dishes, stack them, and tape them together. Place the dishes in an incubator at 37°C for 24 to 48 hours.

## *Part 2—Observations*

**16.** Clean the lab-table surface with disinfectant solution and paper towels.

**17.** After 24 hours, untape the stack of four dishes. *Note: Do not remove the tape from the individual petri dishes.* Observe each dish, and record the number and description of the colonies in the data table below. The growth of off-white colonies indicates the presence of bacteria. The presence of colonies marked with blue indicates that X-gal has been digested. *Note: Do not open sealed petri dishes.*

***Results of Genetic Transformation With pUC8***

| Dish | Conditions | Description of results |
|---|---|---|
| Dish 1 | +pUC8, LB | off-white colony growth |
| Dish 2 | +pUC8, LB + X-gal | off-white colony growth with royal blue circles surrounding the colony |
| Dish 3 | −pUC8, LB | off-white colony growth |
| Dish 4 | −pUC8, LB + X-gal | off-white colony growth |

**18.** Dispose of your materials according to the directions from your teacher.

**19.** Clean up your work area and wash your hands before leaving the lab.

## Analysis

**20.** What is the significance of the blue circles?

The blue circles signify that *E. coli* JM101 was transformed from lac– to lac+. Specifically, the blue circles indicate the presence of a waste product given off when X-gal is metabolized by the enzyme β-galactosidase, which is made by lac+ bacteria.

**21.** Was colony growth the same for each petri dish? What can you conclude from this?

Colony growth should be about the same for each petri dish. Only the presence or absence of a visible waste product should differ. In dishes without X-gal, neither the +pUC8 nor the –pUC8 dishes developed blue rings. In dishes with X-gal, only the bacteria with the plasmid produced blue rings. This shows that the metabolism of X-gal only occurs when bacteria with the plasmid are introduced to X-gal.

**22.** What part of the lab showed that the original *E. coli* bacteria were not able to digest the X-gal?

The bacteria without the plasmid in Dish 4 did not produce blue rings as a waste product when grown on the X-gal-treated dish.

## Conclusions

**23.** In this exploration, you created a new form of *E. coli* by placing a plasmid into a bacterial cell. Did the experimental design work? What applications does transformation have to society?

Answers will vary. New organisms could be developed to help solve problems in society from pollution and waste treatment to creating disease-resistant crops.

**24.** Today, many diabetics take human insulin that is made by bacteria. How do you think bacteria can be made to do this?

The human gene for insulin is inserted into a plasmid. This plasmid is then inserted into bacterial cells. These cells are grown, and the human insulin protein is made during normal metabolism. It is then processed for use by diabetics. This is an example of transformation.

## Extensions

**25.** *Biotechnologists* manage biological systems for human benefit. The career of a biotechnologist has important applications in medicine, food technology, and agriculture. Explore the field of biotechnology, and find out about the training and skills required to become a biotechnologist.

**26.** Do research to find out what laws regulate the manufacturing of new organisms and the impact biotechnology has had on science.

Name ______________________________

Date ______________ Class ______________

HOLT BIOSOURCES
LAB PROGRAM
BIOTECHNOLOGY

# D4 Experimental Design: Genetic Transformation—Antibiotic Resistance

**Prerequisites**

- **Biotechnology D3—Laboratory Techniques: Genetic Transformation of Bacteria** on pages 9–14

**Review**

- cellular respiration
- procedures to induce genetic information
- screening for a characteristic

NIH
National Institutes of Health
Washington, D.C.

March 5, 1998

Caitlin Noonan
Research and Development Division
BioLogical Resources, Inc.
101 Jonas Salk Dr.
Oakwood, MO 65432-1101

Dear Ms. Noonan,

I am a researcher at the National Institutes of Health, or as it is frequently called, the NIH. A recent outbreak of bacterial infections in several European countries has caused us some concern. Because these countries are frequented by American tourists, it is likely that the bacteria causing these infections will soon reach the United States, if they haven't already. The problem with these bacteria is their apparent resistance to certain antibiotics. An outbreak of resistant bacteria could be very difficult to control.

We have been successful in isolating a plasmid from one of these strains of bacteria. The plasmid appears to have a gene for tetracycline resistance. We are asking several research companies to perform experiments with the plasmid so that we can be sure of what we are dealing with. If our suspicions are confirmed, we will take further steps to prepare for a potential outbreak.

I will be contacting you soon to discuss your participation and to provide you with more information. We look forward to your results.

Sincerely,

Lane Conn, M.D.

Lane Conn, M.D.
Director of Research
NIH

BioLogical Resources, Inc. Oakwood, MO 65432-1101

# M E M O R A N D U M

To: Team Leader, Genetic Engineering Dept.

From: Caitlin Noonan, Director of Research and Development

Please review the attached letter. Dr. Conn has sent a sample of the plasmid to which she refers in her letter. Have your research team use this plasmid to genetically transform the RRI strain of *E. coli,* and screen the resulting bacterial colonies for tetracycline resistance. Give this project your top priority. It is important that we finish this project as quickly as possible, but I do not want you to sacrifice quality for speed. I am sure that you understand the importance of a contract with the National Institutes of Health. Take extreme care in providing accurate results.

I recently received word that OSHA will be inspecting us for safety standards over the next few days. I do not expect this to affect your performance; I am certain that you already follow all safety and disposal guidelines as strictly as possible.

## Proposal Checklist

Before you start your work, you must submit a proposal for my approval. **Your proposal must include the following:**

_____ • the **question** you seek to answer

_____ • the **procedure** you will use

_____ • a detailed **data table** for recording observations

_____ • a complete, itemized list of proposed **materials** and **costs** (including use of facilities, labor, and amounts needed)

**Proposal Approval:** ______________________________________

(Supervisor's signature)

## Report Procedures

When you finish your analysis, prepare a report in the form of a business letter to Dr. Conn.
**Your report must include the following:**

_____ • a paragraph describing the **procedure** you followed to complete a genetic transformation of the bacteria

_____ • a complete **data table** pooling data from all research groups

_____ • your **conclusions** about whether the plasmid is tetracycline-resistant

_____ • a detailed **invoice** showing all materials, labor, and the total amount due

## Safety Precautions

- Wear safety goggles, disposable gloves, and a lab apron.
- Wear oven mitts when handling hot objects.
- Glassware is fragile. Notify your teacher promptly of any broken glass or cuts. Do not clean up broken glass or spills unless your teacher tells you to do so.
- Never use electrical equipment around water, or with wet hands or clothing. Never use equipment with frayed cords.
- Wash your hands before leaving the laboratory.

## Disposal Methods

- Dispose of waste material according to instructions from your teacher.
- Place solid, uncontaminated materials in a trash can.
- Place broken glass, unused plasmid, unused nutrient broth, unused bacteria, pipets, cotton swabs, inoculating loops, petri dishes, and other contaminated materials in the separate containers provided.
- Wash reusable materials such as glassware and lab utensils, and return them to the supply area.

## FILE: NIH

**MATERIALS AND COSTS** (Select only what you will need. No refunds.)

**I. Facilities and Equipment Use**

| Item | Rate | Number | Total |
|---|---|---|---|
| facilities | $480.00/day | ______ | ______ |
| personal protective equipment | $10.00/day | ______ | ______ |
| incubator | $20.00/day | ______ | ______ |
| clock or watch with second hand | $10.00/day | ______ | ______ |
| beaker | $5.00/day | ______ | ______ |
| hot plate | $15.00/day | ______ | ______ |
| thermometer | $5.00/day | ______ | ______ |
| test-tube rack | $5.00/day | ______ | ______ |

**II. Labor and Consumables**

| Item | Rate | Number | Total |
|---|---|---|---|
| labor | $40.00/hour | ______ | ______ |
| test plasmid | provided | ______ | ______ |
| stock culture of *E. coli* RRI | $30.00 each | ______ | ______ |
| nutrient broth | $1.00/mL | ______ | ______ |
| petri dish w/ +T agar | $5.00 each | ______ | ______ |
| petri dish w/ −T agar | $5.00 each | ______ | ______ |
| calcium chloride solution | $4.00/mL | ______ | ______ |
| 15 mL plastic tube w/ lid | $2.00 each | ______ | ______ |
| 1 mL sterile plastic pipet | $2.00 each | ______ | ______ |
| sterile graduated pipet | $5.00 each | ______ | ______ |
| pipet bulb | $2.00 each | ______ | ______ |
| plastic-foam tube holder | $5.00 each | ______ | ______ |
| disposable inoculating loops | $0.50 each | ______ | ______ |
| cotton swabs | $0.10 each | ______ | ______ |
| disinfectant solution | $2.00/bottle | ______ | ______ |
| paper towels | $0.10/sheet | ______ | ______ |
| nontoxic permanent marker | $2.00 each | ______ | ______ |
| transparent tape | $0.10/m | ______ | ______ |

**Fines**

| | | | |
|---|---|---|---|
| OSHA safety violation | $2,000.00/incident | ______ | ______ |
| | | **Subtotal** | ______ |
| | | **Profit Margin** | ______ |
| | | **Total Amount Due** | ______ |

Name ______________________________

Date ______________ Class ______________

HOLT BIOSOURCES
LAB PROGRAM
BIOTECHNOLOGY

# D5 Laboratory Techniques: Introduction to Agarose Gel Electrophoresis

## Skill

- conducting agarose gel electrophoresis

## Objectives

**PREPARATION NOTES**
**Time Required:** one 50-minute period

- *Understand* the principles and practices of agarose gel electrophoresis.
- *Determine* the $R_f$ value for each of five dye samples.
- *Demonstrate* that gel electrophoresis can be used to separate a mixture of molecules based on the charge and size of the molecules.

## Materials

**Materials**
Materials for this lab activity can be purchased from WARD'S. See the *Master Materials List* for ordering instructions.

- safety goggles
- lab apron
- agarose gel (2.0%) on gel-casting tray
- electrophoresis system, battery-powered
- microtube rack
- microtube of bromophenol blue
- microtube of crystal violet
- microtube of orange G
- microtube of methyl green
- microtube of xylene cyanol
- microtube of dye mixture
- 10 µL micropipetter
- micropipetter tips (6)
- 250 mL graduated cylinder
- TBE running buffer (1×) (200 mL)
- 250 mL beaker
- 9 V batteries connected in series (5)
- 15 cm metric ruler

## Purpose

You are an intern working in a genetic engineering laboratory. Throughout your internship, you will be assisting genetic engineers in many different experiments with DNA. In some types of DNA analysis, a sample of DNA is broken into fragments, which are then separated according to size. DNA fragments are separated by gel electrophoresis. Before you work with DNA samples, you must first learn this technique. As a training exercise, you will electrophorese five dye samples and a mixture of dyes. Like DNA fragments, the individual dyes in a mixture of dyes separate based on their size and electrical charge.

## Background

**Preparation Tips**
- To prepare 1× TBE running buffer, add 180 mL distilled water to 20 mL of TBE running buffer concentrate (10×). This will make a working buffer solution of 1×. This 200 mL quantity is enough to run one gel.
- Precast gels come with the WARD'S kit. To cast 2.0% agarose gels, refer to product literature supplied with prepared agarose. For this experiment only, be sure to position and

**Gel electrophoresis** is a process that is used to separate mixtures of electrically charged molecules, such as DNA and proteins, on the basis of their size and electrical charge. The process involves passing an electric current through a **gel,** which is a slab made of a jellylike substance. Gels can be made from different substances. The gels that are commonly used to electrophorese (separate through gel electrophoresis) DNA molecules contain **agarose,** which is a sugar that comes from certain types of marine algae. Gel electrophoresis that uses gels containing agarose is called **agarose gel electrophoresis.**

During gel electrophoresis, each sample to be tested is placed in a depression called a **well,** located at one end of the gel. An electric current applied across the gel causes one end of the gel to become negative and the other end to become positive. The electric current causes the samples to migrate through small holes, or pores, in the gel. Molecules that have a negative charge migrate toward the positive electrode. Molecules with a positive charge migrate toward the negative electrode. Small molecules move more easily through the pores in a gel than do large molecules. Therefore, smaller molecules move farther and at a faster rate than larger molecules. After gel electrophoresis, the largest molecules are found closest to their wells, while the smallest molecules are found the farthest away.

## Procedure

align the gel comb in the center of the casting tray.

- Prelabel student microtubes (set of 6 per student team) as follows:

| | |
|---|---|
| Tube 1 | Bromophenol blue |
| Tube 2 | Crystal violet |
| Tube 3 | Orange G |
| Tube 4 | Methyl green |
| Tube 5 | Xylene cyanol |
| Tube 6 | Dye mixture |

- Using a dropper, add 4 to 5 drops of each dye to corresponding student microtubes. Use of a microcentrifuge tube rack is recommended.

**Disposal**

- Dyes can be flushed down the drain with copious amounts of water.
- Dilute 1× TBE buffer in a ratio of 1 part solution to 20 parts water. Flush the diluted solution down the drain with running water.
- Stained agarose gels and other materials used in this lab can be put in the trash.

1. 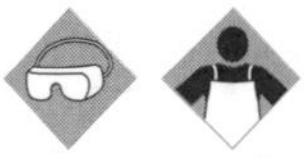 Put on safety goggles and a lab apron.

2. Set a micropipetter to 10 μL, and place a new tip on the end. Open the microtube containing bromophenol blue, and remove 10 μL of the dye. Carefully place the solution in the well in Lane 1 of an agarose gel. To do this, place both elbows on the lab table, lean over the gel, and slowly lower the micropipetter tip into the opening of the well before depressing the plunger. *Note: Do not jab the micropipetter tip through the bottom of the well.*

3. Using a new micropipetter tip for each tube, repeat step 2 for each of the remaining microtubes. Place each dye in a well, according to the following lane assignments:

| | |
|---|---|
| Lane 2 | Crystal violet |
| Lane 3 | Orange G |
| Lane 4 | Methyl green |
| Lane 5 | Xylene cyanol |
| Lane 6 | Dye mixture |

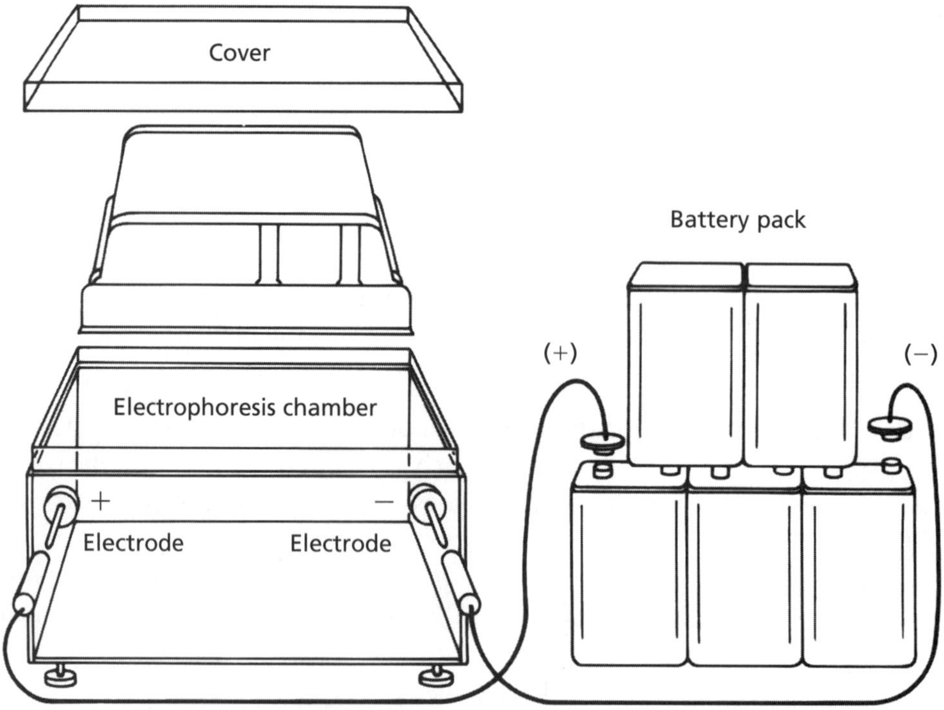


4. Carefully place the agarose gel (still in a gel-casting tray) in the electrophoresis chamber of an electrophoresis apparatus, such as the one shown above.

5. 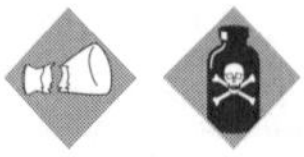 Slowly pour approximately 200 mL of 1× TBE running buffer into a beaker. **CAUTION: Glassware is fragile. Notify your teacher promptly of any broken glass or cuts. Do not clean up broken glass**

**PROCEDURAL NOTES**

**Safety Precautions**

- Discuss all safety symbols and caution statements with students.

**Procedural Tips**

- You may need to demonstrate to students how to use a micropipetter to dispense dye samples into gel wells. Caution students to be very careful not to puncture the gel with the micropipetter tips when dispensing dye samples.
- The dyes on the gel are not fixed and over time (about 24 hours) may diffuse, making the bands less distinct. You need to have your students take the measurements on the same day.
- The gel prepared in this activity does not require staining because the colored dyes will bind onto the gel and form clearly visible bands.

**or spills unless your teacher tells you to do so. CAUTION: If you get a chemical on your skin or clothing, wash it off at the sink while calling to your teacher. If you get a chemical in your eyes, promptly flush it out at the eyewash station while calling to your teacher. Notify your teacher in the event of any chemical spill.** Gently and slowly pour the running buffer from the beaker into one side of the electrophoresis chamber until the gel is completely covered (approximately 1 to 2 mm *above* the top surface of the gel). *Note: Be careful not to overfill the chamber with buffer.*

6. Place the cover on the electrophoresis chamber. Wipe off any spills around the electrophoretic apparatus before doing the next step.

7.  Connect five 9 V alkaline batteries as shown in the diagram on the previous page. **CAUTION: Do not touch both ends of the patch cords or both terminals on the battery pack at the same time.** Connect the red (positive) patch cord to the red terminal on the chamber and the red terminal on the battery pack. Follow the same procedure with the black (negative) patch cord and the black terminals.

8. Observe the migration of the samples along the gel toward the red (positive) electrode and toward the black (negative) electrode.

9. Disconnect the battery pack when the dye bands in Lane 6 are fully separated and when one of the bands is near the end of the gel.

10. Remove the cover from the electrophoresis chamber. Lift the gel tray (containing the gel) from the chamber onto a piece of paper towel. Notch one side of the gel so that you can identify the lanes. *Note: The dyes are not fixed on the gel and over time (several hours) may diffuse into the gel, making the bands less distinct. Complete the next step on the same day.*

11. Use a metric ruler to measure the distance of the dye bands in Lane 6 (in mm) from each of the six sample wells. *Note: Be sure to measure from the center of the well to the center of the band.* Draw an illustration of the dye bands in each gel lane. Make your drawing in the gel diagram provided below.

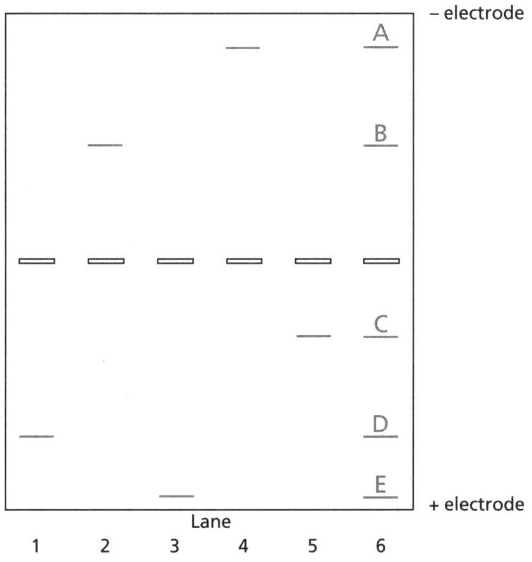

Label the dye bands in Lane 6 of your diagram A–E in order from top to bottom. Dye A will be closest to the negative pole, and Dye E will be closest to the positive pole.

**12.** Record the migration of each sample toward (+) or away from (−) the positive pole in the data table below. Then record the molecular charge (+ or −) of the dye in each band. Also record migration distance (in mm) for each band imaged on the gel.

### *Experimental Data*

| | Migration direction [(+) or (−)] | Molecular charge (+ or −) | Migration distance (mm) | Molecule size (BP)* |
|---|---|---|---|---|
| Lane 1 | (+) | − | 23 | 250 |
| Lane 2 | (−) | + | 15 | 1100 |
| Lane 3 | (+) | − | 31 | 70 |
| Lane 4 | (−) | + | 28 | 110 |
| Lane 5 | (+) | − | 10 | 2800 |
| Lane 6 | | | | |
| Dye A | (−) | + | 28 | 110 |
| Dye B | (−) | + | 15 | 1100 |
| Dye C | (+) | − | 10 | 2800 |
| Dye D | (+) | − | 23 | 250 |
| Dye E | (+) | − | 31 | 70 |

*BP stands for base-pair equivalent.

Remember that agarose gel electrophoresis is commonly used to separate mixtures of DNA molecules based on their length. The length of a DNA molecule is determined by the number of nucleotides that make up each strand of the double-stranded molecule. The two strands are held together by bonds that form between complementary nitrogen bases along each strand. These complementary pairs of nitrogen bases are called **base pairs.**

**13.** You can determine the relative size of each dye's molecules by using the standard curve shown in the graph on the next page. This graph, which was developed by finding the log and antilog of the distance each dye moved, gives the size of the molecules in base-pair equivalents (BP). Find the size of each dye on your gel by determining where the distance the dye moved intersects the standard curve. The number of base-pair equivalents can be read on the *y*-axis.

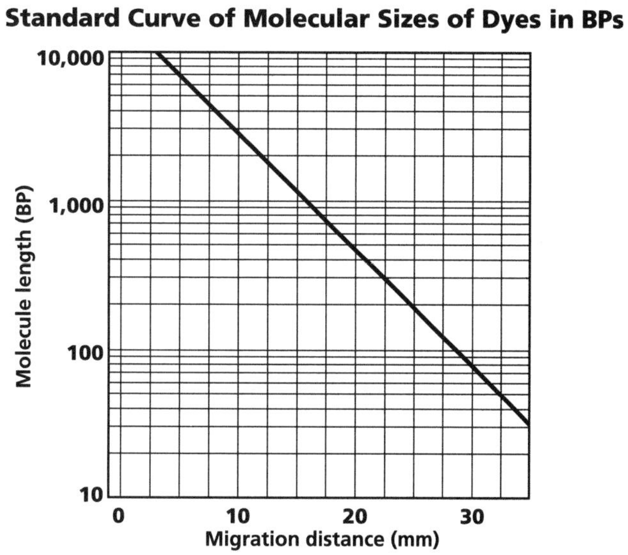


14. Dispose of your materials according to the directions from your teacher.

15. Clean up your work area and wash your hands before leaving the lab.

## Analysis

16. Which dye migrated the farthest distance? the shortest distance?

Orange G, in Lane 3, migrated the farthest distance (31 mm). Xylene cyanol, in Lane 5, migrated the shortest distance (10 mm).

17. Which dye has the largest molecules? Which dye has the smallest molecules? Explain your answer.

Xylene cyanol, the dye in Lane 5, has the largest molecules, and Orange G, the dye in Lane 3, has the smallest. Smaller molecules can migrate farther through the gel and at a faster rate. Larger molecules are found nearer the point of origin.

18. Why do certain dyes migrate toward the cathode and others toward the anode?

The positively charged molecules will migrate toward the cathode, and the negatively charged molecules will migrate toward the anode. As for the dyes used, those having a positive net charge will migrate toward the cathode, and those having a negative net charge will migrate toward the anode.

19. State some ways electrophoresis can separate molecules.

Gel electrophoresis can be used to separate molecules on the basis of size and the electrical charge of the molecule.

## Conclusions

20. What dyes make up the mixture of dyes that you loaded into Lane 6? Identify each dye in the mixture.

The mixture of dyes contains all of the other five dyes used in the activity. Dye A of the mixture is methyl green. Dye B is crystal violet. Dye C is xylene cyanol. Dye D is bromophenol blue. Dye E is orange G.

21. What do you think would happen if you continued to allow electricity to run through the gel?

The dyes would run off the gel.

22. What do you think would happen if you increased the voltage (strength) of the electric current in the gel? decreased the voltage?

Increasing the voltage would cause faster movement. Decreasing the voltage would cause slower movement.

23. DNA and many other substances are colorless and cannot be seen as they move through the gel. How might the dyes that you electrophoresed in this lab be useful during the electrophoresis of DNA?

These dyes can be used as tracking dyes. The crystal violet and methyl green can be used in gel electrophoresis to track positively charged proteins. Bromophenol blue, orange G, and xylene cyanol can be used to track negatively charged DNA.

## Extensions

24. With the help of your teacher, design an experiment to determine what concentration of agarose to use to separate molecules that are very large.

25. Find out how gel electrophoresis is applied in biological laboratories, such as those that investigate the causes of disease and that create recombinant DNA.

Name ______________________________

Date ____________ Class ____________

HOLT BIOSOURCES LAB PROGRAM BIOTECHNOLOGY

# D6 Laboratory Techniques: DNA Fragment Analysis

## Skills

- conducting agarose gel electrophoresis
- calculating $R_f$

## Objectives

- *Use* agarose gel electrophoresis to separate DNA fragments of different sizes.
- *Analyze* DNA fragments to determine their length.

**PREPARATION NOTES**

**Time Required:** three 50-minute periods (first day, run gel; second day, stain gel and destain overnight; third day, analyze results)

**Materials**

Materials for this lab activity can be purchased from WARD'S. See the *Master Materials List* for ordering instructions.

## Materials

- safety goggles
- lab apron
- agarose gel (0.8%) on gel-casting tray
- electrophoresis system, battery-powered
- microtube rack
- samples in labeled microtubes:
  Tube 1 "lambda DNA/*Eco*RI digest"
  Tube 2 "lambda DNA/*Hin*dIII digest"
  Tube 3 "lambda DNA/*Eco*RI/*Hin*dIII double digest"
- 10 μL micropipetter
- micropipetter tips (3)
- 250 mL graduated cylinder
- 1× TBE running buffer (200 mL)
- 250 mL beaker
- 9 V batteries (5)
- "gel handler" spatula
- staining tray
- DNA stain (100 mL per gel)
- metric ruler
- distilled water

## Purpose

You are a graduate student in molecular biology. You have just gotten a summer job working on the Human Genome Project. The goal of this project is to make detailed maps showing the locations of the genes on each of the 24 different types of human chromosomes. To be able to work in the lab, you must become proficient in the following techniques: separating DNA fragments with agarose gel electrophoresis and determining the length of DNA fragments. Your first assignment is to practice these techniques with DNA from the lambda virus.

## Background

**Preparation Tips**

• When using a battery-operated electrophoresis chamber, it will take up to 2 hours for the DNA to separate. If you use three batteries instead of five, the DNA can be separated overnight. The gel can be stained the day after it is electrophoresed. Store the gel wrapped in plastic in a refrigerator.

**Lambda** is a **bacteriophage,** which is a kind of virus that can infect only bacterial cells. Lambda viruses are used in recombinant DNA technology as **vectors,** agents that carry DNA from one organism to another.

Cutting DNA into fragments is the first step in certain types of DNA analysis. The DNA is cut into fragments with a restriction enzyme. A **restriction enzyme,** or **RE,** is an enzyme that recognizes and binds to specific short sequences of base pairs in a DNA molecule and then cleaves, or cuts, the DNA at a specific site within that sequence.

Each type of restriction enzyme cuts DNA at a different base-pair sequence. The recognition sequences at which a restriction enzyme cuts a DNA molecule are relatively short, usually only four to eight base pairs in length. An RE scans the length of a DNA molecule and stops to cut the molecule only at its particular recognition site. The restriction enzyme *Eco*RI, for example, cuts DNA whenever it encounters the base-pair sequence CTTAAG.

Some restriction enzymes cut cleanly through the DNA molecule by cutting both of the complementary strands of the DNA molecule at the same position within the recognition sequence. These enzymes produce a blunt-end cut, as shown in the diagram below. The terms *5'* and *3'* refer to specific carbon atoms in the structure of the sugar in the sugar-phosphate backbone of each strand. Notice that the orientation of the 5' and 3' carbons in the two strands is opposite.

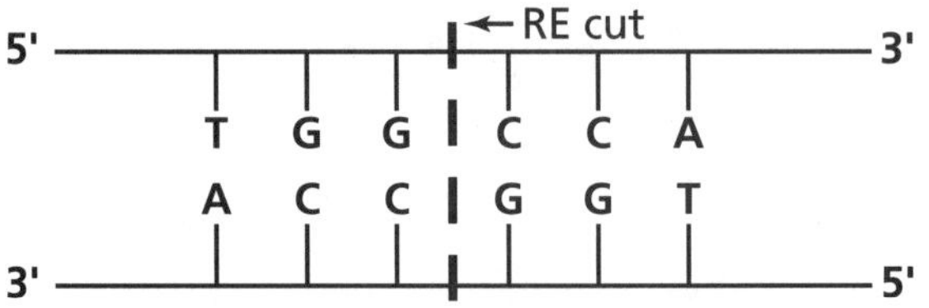


Other restriction enzymes, such as *Eco*RI and *Hin*dIII, cut the complementary strands at a different point within the recognition sequence, resulting in a staggered cut. DNA fragments that result from staggered cuts have single-stranded sections of DNA at their ends. These single-stranded ends are called **sticky ends** because two ends with complementary base sequences will join (stick together) when the complementary bases pair. The diagram below shows how *Hin*dIII cuts a DNA fragment to produce sticky ends.

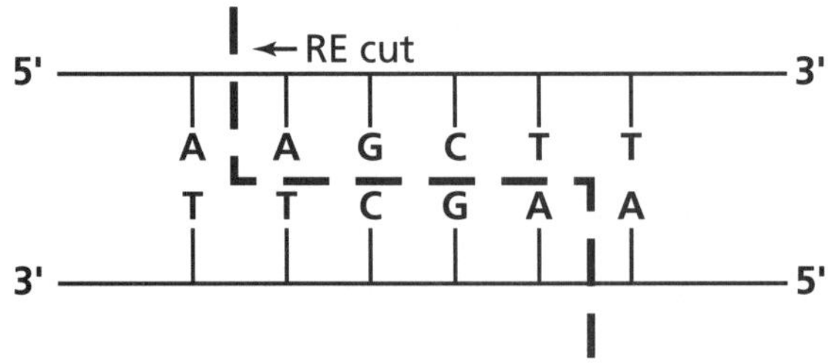


Some of the restriction enzymes that cleave DNA and the positions (base numbers in sequence) at which they cleave lambda DNA are listed in the table below. The positions in the table refer to the 5' base of the recognition sequence. Site position(s) read from left to right (5'→3' position) along the lambda DNA.

***Table 1 Restriction Enzymes and Lambda DNA Cleavage Sites***

| Enzyme | No. of sites | Base position of cleavage sites | | | | | | |
|---|---|---|---|---|---|---|---|---|
| | | 1 | 2 | 3 | 4 | 5 | 6 | 7 |
| *Bam*HI | 5 | 5505 | 22346 | 27972 | 34499 | 41732 | | |
| *Eco*RI | 5 | 21226 | 26104 | 31747 | 39168 | 44972 | | |
| *Hin*dIII | 7 | 23130 | 25157 | 27479 | 36895 | 37459 | 37584 | 44141 |

The fragments of DNA molecules that result when a sample of DNA is cut with restriction enzymes can be separated by agarose gel electrophoresis. During gel electrophoresis, the DNA cut by one or more restriction enzymes is loaded into a well of an agarose gel. The wells are placed near the negative electrode in a gel electrophoresis chamber. When an electric current passes through the gel, the ends of the gel become electrically charged. The DNA molecules, which are negatively charged, migrate toward the positive end of the gel.

- To prepare 1× TBE running buffer, add 180 mL distilled water to 20 mL of TBE running buffer concentrate (10×). This will make a working buffer solution of 1×. This 200 mL quantity is enough to run one gel.
- To cast 0.8% agarose gels, refer to the product literature supplied with prepared agarose. Each student team will run three sample lanes: lambda DNA/*Eco*RI, lambda DNA/*Hin*dIII, lambda DNA/*Eco*RI/*Hin*dIII double digest.
- Prelabel student microtubes (set of three per student team) as follows:
  - Tube 1—lambda DNA/*Eco*RI digest
  - Tube 2—lambda DNA/*Hind* III digest
  - Tube 3—lambda DNA/*Eco*RI /*Hind* III double digest
- Set up two dispensing stations, rather than passing out microtubes of DNA to students.
- Optional: Heat Tube 2 at 65°C for 5 minutes. Heating helps band imaging. *Note: Use the gray sample container in which your DNA was packaged from WARD'S as a "micro water bath."* Place Tube 2 into the container and pour water (90°C) up to the dividers. Close the cover. Allow to stand for 5 minutes. Average incubation temperature will be 65°C.

**Disposal**

- DNA stain and samples of DNA can be poured down the drain with copious amounts of water.
- Dilute 1× TBE buffer in a ratio of 1 part solution to 20 parts water. Flush the diluted solution down the drain with running water.

## Procedure

• Stained agarose gels and other materials used in this lab can be put in the trash.

**PROCEDURAL NOTES**

**Safety Precautions**

• Discuss all safety symbols and caution statements with students.

• Remind students to notify you immediately of any chemical spills. Also caution them to never taste, touch, or smell any substance or bring it close to their eyes.

**Procedural Tips**

• Use the "gel handler" spatula to remove the gel from the staining tray. Wrap the gel in clear plastic wrap, and store it in the refrigerator. Gel bands will retain the stain for about one month before they start to fade. Fading is due to the oxidation of DNA because DNA is not fixed in the gel matrix.

1. Put on safety goggles and a lab apron.

### *Part 1—Separating DNA Fragments*

2. Set a micropipetter to 10 µL, and place a new tip on the end. Open the microtube containing lambda DNA/*Eco*RI digest (Tube 1) and remove 10 µL of the solution. Carefully place the solution into the well in Lane 1 of an agarose gel in a gel-casting tray. To do this, place both elbows on the lab table, lean over the gel, and slowly lower the micropipetter tip into the opening of the well before depressing the plunger. *Note: Do not jab the micropipetter tip into the gel.*

3. Using a new micropipetter tip for each tube, repeat step 3 for each of the remaining microtubes:
   Lane 2 (Tube 2—lambda DNA/*Hin*dIII digest)
   Lane 3 (Tube 3—lambda DNA/*Eco*RI/*Hin*dIII double digest)

4. Carefully place your loaded agarose gel (still in the gel-casting tray) in the electrophoresis chamber of an electrophoresis apparatus, such as the one shown below. Orient the gel so that the wells are closest to the black connector, or negative electrode.

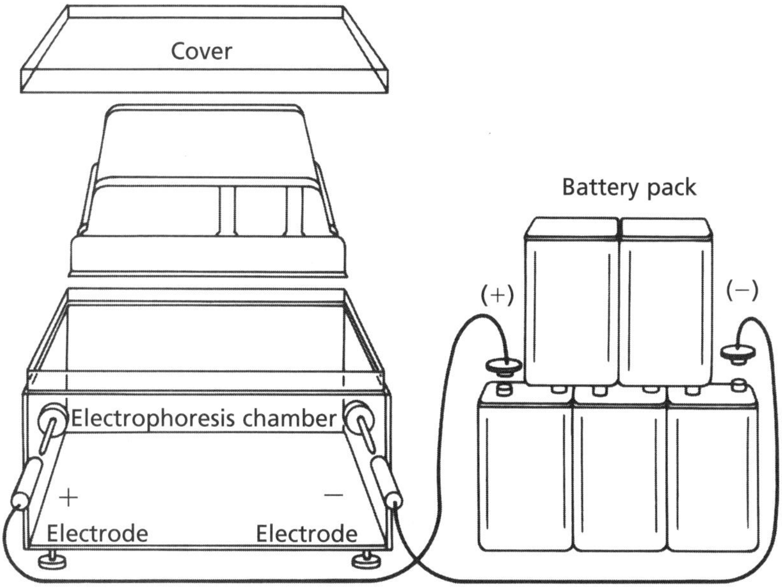


5. Slowly pour approximately 200 mL of 1× TBE running buffer into a beaker. **CAUTION: Glassware is fragile. Notify your teacher promptly of any broken glass or cuts. Do not clean up broken glass or spills unless your teacher tells you to do so. If you get a chemical on your skin or clothing, wash it off at the sink while calling to your teacher. If you get a chemical in your eyes, promptly flush it out at the eyewash station while calling to your teacher. Notify your teacher in the event of any chemical spill.** Gently and slowly pour the running buffer from the beaker into one side of the electrophoresis chamber until the gel is completely covered (1 to 2 mm *above* the top surface of the gel). *Note: Be careful not to overfill the chamber with buffer.*

6. Place the cover on the electrophoresis chamber. Wipe off any spills around the electrophoretic apparatus before doing the next step.

7. Connect five 9 V alkaline batteries as shown in the figure on the preceding page. **CAUTION: Do not touch both ends of the patch cords or both terminals on the battery pack at the same time.** Connect the red (positive) patch cord to the red terminal on the chamber and the red terminal on the battery pack. Follow the same procedure with the black (negative) patch cord and the black terminals.

8. The samples you placed in the wells contain a tracking dye. Observe the migration of the tracking dye along the gel toward the red (positive) electrode.

9. Disconnect the battery pack when the tracking dye has run near the end of the gel.

10. Remove the cover from the electrophoresis chamber. Lift the gel tray (containing the gel) from the cell, and place the gel in the staining tray. Do this by *gently* pushing the gel off the tray using a "gel handler" spatula. Notch one side of the gel so that you can identify the lanes.

### Part 2—Staining a Gel

11. To stain the gel, pour approximately 100 mL of DNA stain into the staining tray until the gel is completely covered. Do not pour stain directly onto the gel. Cover and label the staining tray. Staining will be complete in 2–3 hours. **CAUTION: DNA stain will stain your skin and clothing. Promptly wash off spills to minimize staining.**

12. When the gel is stained, carefully pour the remaining stain directly into a sink. Flush down the drain with water. *Note: Do not allow the gel to move against the corner of the staining tray. The gel must remain flat; if it breaks, it will be useless.*

13. To destain the gel, add enough distilled water to the staining tray to cover the gel. *Note: Do not pour water directly onto the gel. Pour water to one side of the gel.* Let the gel sit overnight (or at least 18–24 hours). After destaining the gel, the bands of DNA will appear as purple lines against a light, almost clear background.

14. Dispose of your materials according to the directions from your teacher.

15. Clean up your work area and wash your hands before leaving the lab.

### Part 3—DNA Fragment Analysis

The stained gel you produced in the steps above can be analyzed to determine the lengths of the DNA fragments. The length of DNA fragments is frequently given in nucleotide base pairs (bp) for smaller fragments and kilobase pairs (kbp) for larger ones. Smaller DNA fragments migrate through a gel faster than larger ones. The largest fragments (those with the most base pairs) remain closest to the well after electrophoresis is complete. The smallest fragments (those with the fewest base pairs) move farthest away from the well. Therefore, DNA fragments spread out in a gel lane in order from the highest to the lowest number of base pairs.

The distance a fragment moves through a gel is used to determine the fragment's length. Each fragment has a "relative mobility," or $\boldsymbol{R_f}$. The $R_f$ value is calculated by using the following formula:

$$R_f = \frac{\text{distance the DNA fragment has migrated from the origin (gel well)}}{\text{distance from the origin to the reference point (tracking dye)}}$$

Under a given set of electrophoretic conditions (i.e., pH, voltage, time, gel type and concentration, etc.), the $R_f$ of a DNA fragment is standard. Thus, samples containing DNA fragments of unknown length are usually electrophoresed with a marker sample, which contains fragments of known lengths. The length of an unknown DNA fragment can then be determined by comparing its $R_f$ with those of the fragments in the DNA marker sample. One way you can do this is to use a **standard curve,** which is constructed by plotting migration distance in millimeters versus the log of fragment length.

**16.** Use a metric ruler to measure the distance in millimeters from the well in each lane to the tracking dye and DNA bands in the same lane. *Note: Measure from the middle of a well to the middle of a band.* As you measure the distance each DNA band moved, record this measurement under "Migration distance" in the appropriate data table on the following pages. Draw the position of the DNA bands in each lane of your gel on the "gel" diagram below. *Note: Also draw the position of the wells, based on their position on your gel.*

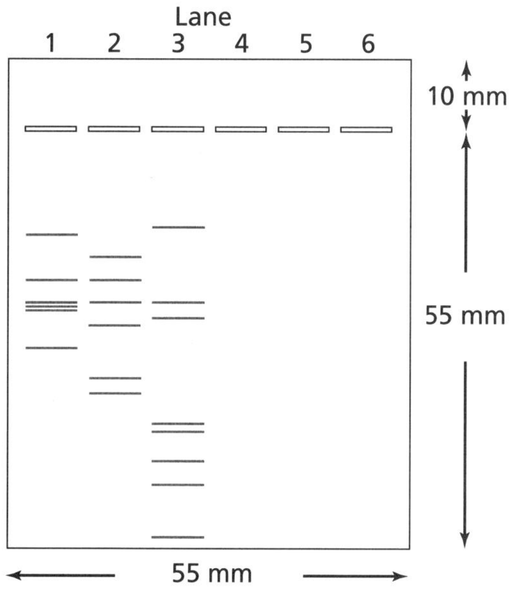


**17.** Compute the log of the size of each of the DNA marker sample's fragments using the data in the table on the next page. *Note: Fragment 1 is the fragment that is closest to the well.*

***Table 2 Data for DNA Marker Sample (Lambda DNA/HindIII Digest)***

| Fragment number | Fragment length (bp) | Log | Migration distance (mm) | $R_f$ |
|---|---|---|---|---|
| 1 | 23,130 | 4.364 | 17 | 0.28 |
| 2 | 9,416 | 3.974 | 20 | 0.34 |
| 3 | 6,557 | 3.817 | 23 | 0.38 |
| 4 | 4,361 | 3.640 | 26 | 0.43 |
| 5 | 2,322 | 3.366 | 33 | 0.55 |
| 6 | 2,027 | 3.307 | 35 | 0.58 |
| 7 | 564 | 2.751 | not imaged | ———— |
| 8 | 125 | 2.097 | not imaged | ———— |

**18.** Calculate the $R_f$ value for each fragment in the DNA marker sample (Lambda DNA/*Hin*dIII digest). Record these values in the table above.

**19.** Using the grid below, construct a standard curve for the DNA marker sample by plotting the $R_f$ value of each fragment versus the log of its length.

***Standard Curve for the Lambda DNA/HindIII Marker Sample***

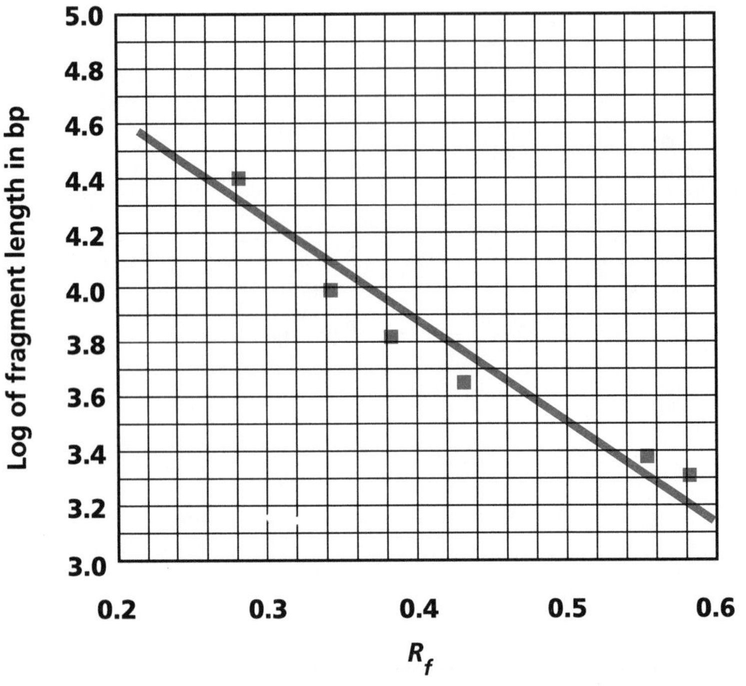

**20.** Calculate the $R_f$ value for each fragment in the *Eco*RI and *Eco*RI/*Hin*dIII digests. Record these values in the data table below.

***Table 3 Data for* Eco*RI and* Eco*RI/*Hind*III Digests***

| *Eco*RI | | | | | *Eco*RI/*Hind* III | | | | |
|---|---|---|---|---|---|---|---|---|---|
| Fragment number | Migration distance (mm) | $R_f$ | Log | Fragment length (bp) | Fragment number | Migration distance (mm) | $R_f$ | Log | Fragment length (bp) |
| 1 | 14.0 | 0.23 | 4.32 | 21,226 | 1 | 13.0 | 0.22 | 4.32 | 21,226 |
| 2 | 20.0 | 0.33 | 3.87 | 7,421 | 2 | 23.0 | 0.38 | 3.71 | 5,148 |
| 3 | 23.0 | 0.38 | 3.76 | 5,804 | 3 | 25.0 | 0.42 | 3.70 | 4,973 |
| 4 | 23.5 | 0.39 | 3.75 | 5,643 | 4 | 39.0 | 0.65 | 3.63 | 4,268 |
| 5 | 24.0 | 0.40 | 3.69 | 4,878 | 5 | 40.0 | 0.67 | 3.55 | 3,530 |
| 6 | 29.0 | 0.48 | 3.55 | 3,530 | 6 | 44.0 | 0.73 | 3.31 | 2,037 |
| | | | | | 7 | 47.0 | 0.78 | 3.28 | 1,894 |
| | | | | | 8 | 54.0 | 0.90 | 3.20 | 1,584 |
| | | | | | 9 | 57.0 | 0.95 | 3.14 | 1,375 |

**21.** Use the standard curve you plotted in step 19 and the $R_f$ values you calculated in step 20 to find the log of the length in base pairs (bp) of each fragment in the *Eco*RI and *Eco*RI/*Hin*dIII digests. Record the logs in the data table above. The antilog of the log for each fragment is its number of base pairs.

## Analysis

**22.** Where on your gel are the fragments with the largest $R_f$ values?

The fragments with the largest $R_f$ values are located farthest from the wells.

**23.** Was the DNA in any of the bands darker than in others? What could account for this?

Answers may vary. Some bands could appear darker because fragments similar in size may appear as one band.

## Conclusion

**24.** In this lab, DNA from the same type of virus was cut with different restriction enzymes. What results would you expect from the gel electrophoresis of DNA from different organisms but cut by the same restriction enzyme?

If the DNA from different organisms (or even different individuals of the same species) were cut with the same restriction enzyme, the fragments would be different because the exact sequence of base pairs in the DNA would be different.

**25.** Would you expect different individuals to have the same or different patterns of DNA bands? Justify your answer.

The banding patterns that result when DNA from different individuals is cut by restriction enzymes and separated by gel electrophoresis would be different because everyone, except identical twins, has a different combination of genes. Therefore, the base sequences of their DNA will not be exactly the same.

## Extensions

**26.** Restriction endonucleases are frequently named after the genus of the organism from which the enzyme was isolated. Find out how the names are developed, and be able to identify the various parts of a name.

**27.** Find out how restriction enzymes are being used to develop recombinant DNA for the production of vaccines and antibiotics that are used to fight human diseases.

Name ______________________________

Date ______________ Class ______________

HOLT BIOSOURCES
LAB PROGRAM
BIOTECHNOLOGY

# D7 Laboratory Techniques: DNA Ligation

## Skills

- conducting agarose gel electrophoresis

## Objectives

- *Perform* a DNA ligation.
- *Use* agarose gel electrophoresis to separate DNA fragments of different sizes.
- *Analyze* DNA fragments to determine their length.

**PREPARATION NOTES**

**Time Required:** two 50-minute periods (first day, run gel; second day, stain gel and destain overnight)

**Materials**

Materials for this lab activity can be purchased from WARD'S. See the *Master Materials List* for ordering instructions.

**Preparation Tips**

- When you are using a battery-operated electrophoresis chamber, it will take up to 2 hours for the DNA to separate. If you use three batteries instead of five, the DNA can be separated overnight. The gel can be stained the day after it is electrophoresed. Store the gel wrapped in plastic in a refrigerator.

## Materials

- safety goggles
- lab apron
- microtube rack
- samples in labeled microtubes:
  Tube 1 "lambda DNA/*Hind* III" (marker)
  Tube 2 "lambda DNA/*Eco*RI" (control)
  Tube 3 "lambda DNA/*Eco*RI" (ligated)
- 1–5 μL micropipets (3)
- 1× ligation buffer (2 μL)
- T4 DNA ligase (3 μL)
- watch or clock with second hand
- ice-water bath
- thermometer
- hot-water bath
- loading (tracking) dye
- 10 μL micropipetter
- micropipetter tips (3)
- agarose gel (0.8%) on gel-casting tray
- 250 mL graduated cylinder
- 1× TBE running buffer 1× (200 mL)
- 250 mL beaker
- electrophoresis system, battery-powered
- 9 V batteries (5)
- "gel handler" spatula
- DNA stain (100 mL per gel)
- staining tray
- distilled water
- metric ruler
- calculator

## Purpose

You are a laboratory technician working for a company that does genetic engineering. Three batches of your DNA ligase enzymes have not been working properly. The laboratory director wants to know if the remaining supplies of enzymes are good. You know some tubes of the enzymes were left out over a hot weekend. You decide to ligate, or join, fragments of lambda DNA that have been cut with the restriction enzyme *Eco*RI. You will test your results by using agarose gel electrophoresis and then comparing the banding pattern of the ligated DNA with the same DNA prior to ligation.

## Background

- To prepare 1× TBE running buffer, add 180 mL of distilled water to 20 mL of TBE running buffer concentrate (10×). This will make a working buffer solution of 1×. This 200 mL quantity is enough to run one gel.

**DNA ligation,** joining fragments of DNA, is a procedure widely used for creating molecules of recombinant DNA. **Recombinant DNA** is a DNA molecule formed when fragments of DNA from two or more different organisms are joined together. Recombinant DNA is then inserted into a vector, viral or bacterial DNA, and allowed to infect target cells. The infected cells are allowed to reproduce a large number of identical cells that contain the recombinant DNA.

**DNA ligases** are enzymes that join fragments of DNA. These enzymes act as catalysts, joining two strands of DNA. One of the most thoroughly investigated ligases is T4 DNA ligase from the bacteriophage T4. Many other types of DNA

- To cast 0.8% agarose gels, refer to the product literature supplied with prepared agarose. Each student team will run three sample lanes: lambda DNA/*Hin*dIII, lambda DNA/*Eco*RI, and lambda DNA/*Eco*RI (ligated).
- Because the amounts of the DNA samples are so small, you may want to set up a dispensing station at which students can obtain their samples.
- Any unused DNA sample should be stored at −3°C.
- Heat Tube 1 (DNA marker) at 65°C for 5 minutes. Heating helps band imaging.

## Procedure

- Set up an ice-water bath (16°C) and a hot-water bath (65°C). You may use a beaker to heat the water to 65°C on a hot plate if a water bath is not available.
- Set up a dispensing station for the T4 DNA ligase, T4 ligase buffer, and loading dye for the students. You will need 1–5 μL micropipetters.

### Disposal

- Dyes and stained samples of DNA can be poured down the drain with copious amounts of water.
- Dilute 1× TBE buffer and 1× ligation buffer in a ratio of 1 part solution to 20 parts water. Flush the diluted solutions down the drain with running water.

ligases have been found in both plant and animal cells, including some mammalian cells. The construction of the first recombinant DNA molecule in the early 1970s marked the birth of the field of **genetic engineering**—moving genes from the chromosomes of one organism to those of another.

*Eco*RI is a **restriction enzyme** (RE), which is an enzyme that scans the length of a DNA molecule and cuts it only at a particular site. *Eco*RI cuts the complementary strands of a DNA molecule at a different point within the recognition sequence, resulting in a staggered cut, as shown in the diagram below.

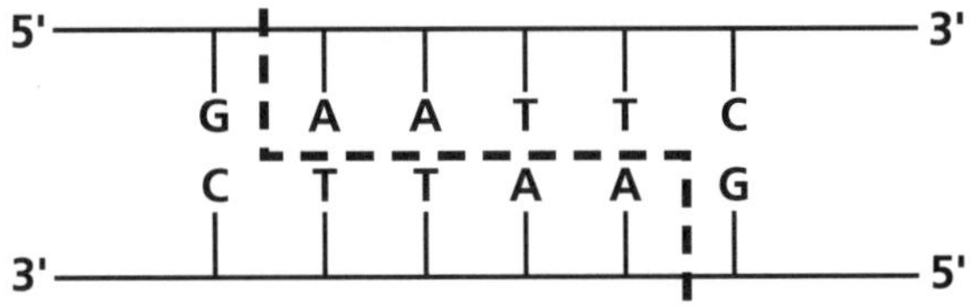


A staggered cut in a DNA molecule exposes single-stranded regions of the molecule. These single-stranded regions, called **sticky ends,** are useful in making recombinant DNA molecules. The sticky ends allow complementary regions in two DNA fragments to recognize one another and join during ligation.

### *Part 1—DNA Ligation*

1. Put on safety goggles and a lab apron.

2. Obtain one each of Tubes 1, 2, and 3. *Note: The sample marked "lambda DNA/*HindIII*" will serve as a marker, which you will use to construct a standard curve that can be used to find the length of the DNA molecules in the other two samples.* Use clean micropipets to add 2 μL of 1× ligation buffer and 3 μL of T4 DNA ligase to Tube 3, which contains 10 μL of lambda DNA that has been cut with *Eco*RI.

3. Incubate Tube 3 at 16°C for 20 minutes in an ice-water bath. At this temperature, the T4 DNA ligase catalyzes the ligation of greater than 95% of the lambda DNA fragments.

4. To inactivate the enzyme, incubate Tube 3 in a 65°C hot-water bath for 10 minutes.

5. Remove Tube 3 from the hot-water bath, and add 2 μL of loading dye. Your sample is now ready for gel electrophoresis. To test whether the ligation procedure was successful, electrophorese samples from each of Tubes 1, 2, and 3.

### *Part 2—Loading and Running a Gel*

6. Set a micropipetter to 10 μL, and place a new tip on the end. Open Tube 1 containing lambda DNA/*Hin*dIII, and remove 10μL of the solution. Carefully place the solution in the well in Lane 1 of an agarose gel in a gel-casting tray. To do this, place both elbows on the lab table, lean over the gel, and slowly lower the micropipetter tip into the opening of the well before depressing the plunger. *Note: Do not jab the micropipetter tip through the bottom of the well.*

• Stained agarose gels and other materials used in this lab can be put in the trash.

**PROCEDURAL NOTES**

**Safety Precautions**

• Discuss all safety symbols and caution statements with students.

• Remind students to notify you immediately of any chemical spills. Also caution them to never taste, touch, or smell any substance or bring it close to their eyes.

**Procedural Tips**

• The plot of lambda DNA/*Hin*dIII is shown in the student graph on page 37. Individual student results will vary from the sample data.

• Note that the distances from individual wells to the end of the gel are not the same. Therefore, students should calculate $R_f$s based on well distances measured for individual lanes.

• Have students use a "gel handler" spatula to remove the gel from the staining tray. Wrap the gel in clear plastic, and store it in the refrigerator. Gel bands will retain the stain for about one month before they start to fade. Fading is due to the oxidation of DNA because DNA is not fixed in the gel matrix.

**7.** Using a new micropipetter tip for each tube, repeat step 6 for each of the remaining microtubes:

Lane 2 (Tube 2) lambda DNA/*Eco*RI (control)

Lane 3 (Tube 3) lambda DNA/*Eco*RI (ligated)

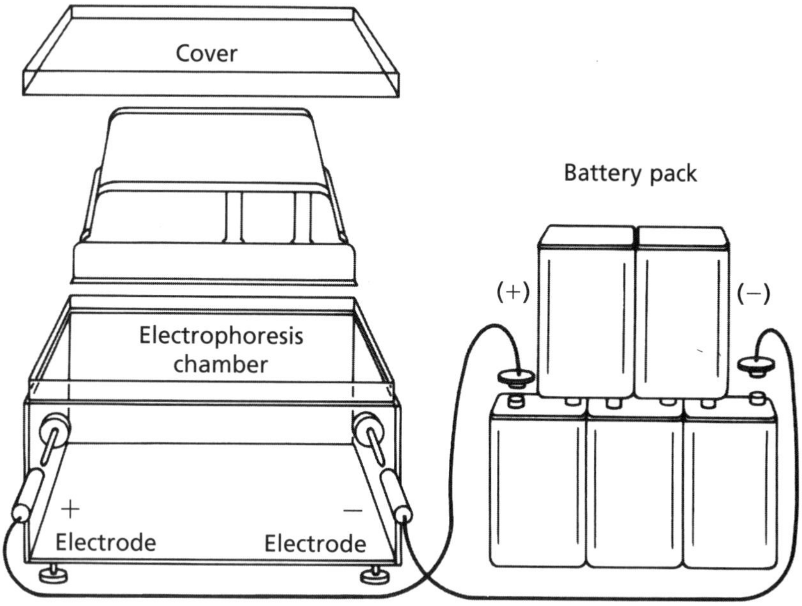


**8.** Carefully place your loaded agarose gel (still in the gel-casting tray) in the electrophoresis chamber of an electrophoresis apparatus, such as the one shown above. Orient the gel so that the wells are closest to the black connector, or negative electrode.

**9.** Slowly pour approximately 200 mL of 1× TBE running buffer into a beaker. **CAUTION: Glassware is fragile. Notify your teacher promptly of any broken glass or cuts. Do not clean up broken glass or spills unless your teacher tells you to do so. If you get a chemical on your skin or clothing, wash it off at the sink while calling to your teacher. If you get a chemical in your eyes, promptly flush it out at the eyewash station while calling to your teacher. Notify your teacher in the event of any chemical spill.** Gently and slowly pour the running buffer from the beaker into one side of the electrophoresis chamber until the gel is completely covered (1 to 2 mm above the top surface of the gel). *Note: Be careful not to overfill the chamber with buffer.*

**10.** Place the cover over the electrophoresis chamber. Wipe off any spills around the electrophoretic apparatus before doing the next step.

**11.** Connect five 9 V alkaline batteries as shown in the diagram above. **CAUTION: Do not touch both ends of the patch cords or both terminals on the battery pack at the same time.** Connect the red (positive) patch cord to both the red electrode terminal on the chamber and the red terminal on the battery pack. Follow the same procedure with the black (negative) patch cord.

12. Observe the migration of the tracking, or loading, dye down the gel toward the red (positive) electrode.

13. Disconnect the battery pack when the tracking dye has reached the end of the gel.

14. Remove the cover from the electrophoresis chamber. Lift the gel tray (containing the gel) from the cell and place the gel in the staining tray. Do this by *gently* pushing the gel off the tray using a "gel handler" spatula.

### *Part 3—Staining a Gel*

15. To stain the gel, pour approximately 100 mL of DNA stain into the staining tray until the gel is completely covered. Do not pour stain directly onto the gel. **CAUTION: DNA stain will stain your skin and clothing. Promptly wash off spills to minimize staining.** Cover and label the staining tray. Staining will be complete in 2 to 3 hours.

16. When the gel is stained, carefully pour the remaining stain directly into a sink. Flush down the drain with water. *Note: Do not allow the gel to move against the corner of the staining tray. The gel must remain flat; if it breaks, it will be useless.*

17. To destain the gel, add enough distilled water to the staining tray to cover the gel. *Note: Do not pour water directly onto the gel. Pour water to one side of the gel.* Let the gel sit overnight (or at least 18–24 hours). After destaining the gel, the bands of DNA will appear as purple lines against a light, almost clear background.

18. Dispose of your materials according to the directions from your teacher.

19. Clean up your work area and wash your hands before leaving the lab.

## Analysis

20. Use a metric ruler to measure the distance in millimeters from the well in each lane to the tracking dye and DNA bands in the same lane. *Note: Measure from the middle of a well to the middle of a band.* As you measure the distance each DNA band moved, record this measurement under "Migration distance" in the appropriate data table on the following pages. Draw the position of the DNA bands in each lane of your gel on the "gel" diagram below.

**21.** Compute the log of the size of each of the Lambda DNA/*Hind*III marker sample's fragments using the data in the table below. *Note: Fragment 1 is the fragment that is closest to the well.*

***Data for DNA Marker Sample (Lambda DNA/HindIII Digest)***

| Fragment number | Fragment length | Log | Distance migrated (mm) | $R_f$ |
|---|---|---|---|---|
| 1 | 23,130 | 4.364 | 17 | 0.28 |
| 2 | 9,416 | 3.974 | 20 | 0.34 |
| 3 | 6,557 | 3.817 | 23 | 0.38 |
| 4 | 4,361 | 3.640 | 26 | 0.43 |
| 5 | 2,322 | 3.366 | 33 | 0.55 |
| 6 | 2,027 | 3.307 | 35 | 0.58 |
| 7 | 564 | 2.751 | not imaged | ——— |
| 8 | 125 | 2.097 | not imaged | ——— |

**22.** In the space below, calculate the $R_f$ value for each of the DNA marker sample's fragments by using the following formula:

$$R_f = \frac{\text{distance the DNA fragment has migrated from the origin (gel well)}}{\text{distance from the origin to the reference point (tracking dye)}}$$

Record these values in the table above.

**23.** Using the grid below, construct a standard curve for the DNA marker sample by plotting the $R_f$ value of each fragment versus the log of its length.

***Standard Curve for the Lambda DNA/Hind*III *Marker Sample***

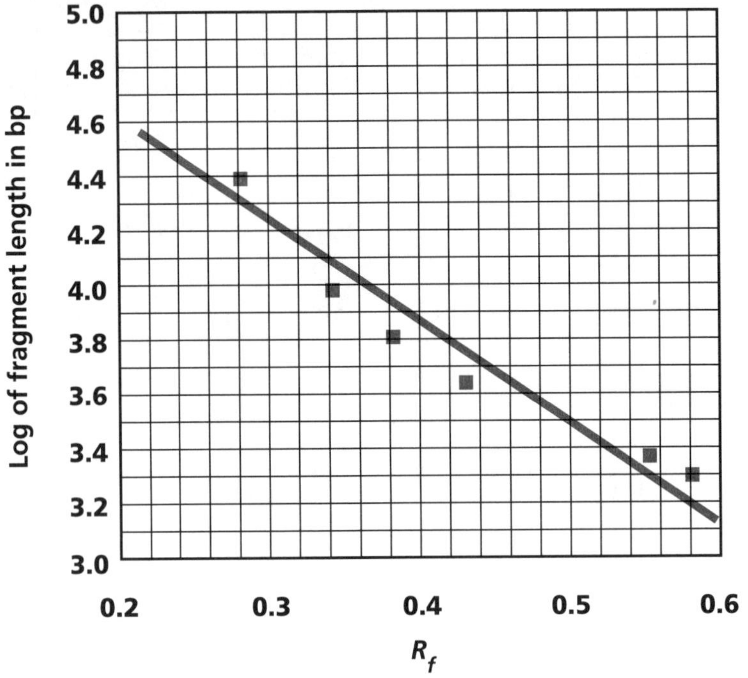


**24.** In the space below, calculate the $R_f$ value for each fragment in the *Eco*RI (control) and *Eco*RI (ligated) samples. Record these values in the data table on the next page.

### *Data for EcoRI (Control) and EcoRI (Ligated)*

| *Eco*RI (control) | | | | | *Eco*RI (ligated) | | | | |
|---|---|---|---|---|---|---|---|---|---|
| Fragment number | Migration distance (mm) | $R_f$ | Log | Fragment length (bp) | Fragment number | Migration distance (mm) | $R_f$ | Log | Fragment length (bp) |
| 1 | 14.0 | 0.23 | 4.32 | 21,226 | 1 | 10.6 | 0.18 | 4.69 | 48,502 |
| 2 | 20.0 | 0.33 | 3.87 | 7,421 | | | | | |
| 3 | 23.0 | 0.38 | 3.76 | 5,804 | | | | | |
| 4 | 23.5 | 0.39 | 3.75 | 5,643 | | | | | |
| 5 | 24.0 | 0.40 | 3.69 | 4,878 | | | | | |
| 6 | 29.0 | 0.48 | 3.55 | 3,530 | | | | | |

**25.** Use the standard curve you plotted in step 23 and the $R_f$ values you calculated in step 24 to find the log of the length in base pairs (bp) of each fragment in the *Eco*RI (control) and *Eco*RI (ligated) samples. Record the logs in the data table above. The antilog of the log for each fragment is its number of base pairs.

## Conclusions

**26.** Based on your gel analysis, was your ligation reaction successful? Explain your answer.

Yes, because there was only one fragment in the ligated sample and it migrated more slowly (traveled less distance in the same time) than the control, which consisted of several fragments. This means the fragment in the ligated sample is larger (and has more base pairs).

**27.** What is your report concerning the condition of your company's supply of ligation enzyme?

The enzyme is good because it worked.

**28.** Compare the number of base pairs in the bands in Lane 2 with the number of base pairs in the band in Lane 3. What conclusion can you draw from this comparison?

Students should notice one fragment (with more base pairs) of DNA in Lane 3 and six bands in Lane 2. The total number of base pairs in the DNA in Lane 3 should equal the sum of the number of base pairs in the six DNA fragments in Lane 2. Since one band was formed by the ligated DNA in lane 3, the separate DNA fragments must have been joined.

## Extensions

**28.** Find out the importance of the Ti plasmid in the formation of recombinant DNA in plants.

**29.** When genetic engineering became a reality in the early 1970s, biologists were concerned about the potential dangers of producing new strains of pathogenic and lethal microorganisms. Find out what strict protocols are in place today to prevent such misuse of recombinant DNA technology.

Name ______________________

Date ____________ Class ____________

HOLT BIOSOURCES
LAB PROGRAM
BIOTECHNOLOGY

# D8 Experimental Design: Comparing DNA Samples

**Prerequisites**

- **Biotechnology D5—Laboratory Techniques: Introduction to Gel Electrophoresis** on pages 19–24
- **Biotechnology D6—Laboratory Techniques: DNA Fragment Analysis** on pages 25–32

**Review**

- use of restriction enzymes to cut DNA
- procedure for analyzing a gel

*Williams & Associates*

Hollywood, California

December 10, 1997

Caitlin Noonan
Research and Development Division
BioLogical Resources, Inc.
101 Jonas Salk Dr.
Oakwood, MO 65432-1101

Dear Ms. Noonan,

As a lawyer here in Hollywood, I represent many actors and actresses. Following the recent death of actress Jessica Coleman, I became the executor of Ms. Coleman's will. A few days later, a woman by the name of Ms. Wilson arrived at my office claiming to be Ms. Coleman's long-lost identical twin. I read through the will, but found no reference specific to any individual. Instead, the will simply states that the estate is to be divided equally among Ms. Coleman's closest living relatives. Until now, it was presumed that the entire estate would go to her only daughter. However, Ms. Wilson has provided a great deal of evidence to support her claim, and blood tests have not ruled out the possibility that the woman could be Ms. Coleman's twin.

We have blood samples from Ms. Coleman and Ms. Wilson. We are requesting that you perform the necessary DNA tests to determine whether Ms. Wilson could be the twin of the deceased. Please contact me as soon as possible.

Sincerely,

Matthew Williams

Matthew Williams
Attorney at Law
Williams & Associates

## BioLogical Resources, Inc. Oakwood, MO 65432-1101

### M E M O R A N D U M

To: Team Leader, Genetics Dept.

From: Caitlin Noonan, Director of Research and Development

The case described in the attached letter sounds like a challenging one. Our research teams will be using the DNA fingerprinting procedure to compare the DNA of Ms. Coleman and Ms. Wilson. I have had the Biochemistry department extract DNA from each of the blood samples supplied by Mr. Williams, and they have also cut the DNA samples using restriction enzymes.

I want your research teams to separate the DNA fragments that resulted when restriction enzymes cut the DNA. These samples, which are ready for the gel electrophoresis procedure except for the addition of loading dye, have been delivered to your lab. You will also be provided with a marker sample containing RFLPs of known lengths. The lengths of these fragments, from fragment 1 to fragment 8, are as follows: 23,130; 9,416; 6,557; 4,361; 2,322; 2,027; 564; and 125. The logs of these fragments are 4.364, 3.974, 3.817, 3.640, 3.366, 3.307, 2.751, 2.097, respectively. Use the marker sample to determine the lengths of the RFLPs in the test samples. After your team has completed the gel electrophoresis and DNA fragment analysis, the Biochemistry team will complete the DNA fingerprints for each test sample.

Please be very thorough in recording your procedure. This may become important if the case ever develops into a law suit and we are asked to explain our procedure to a jury.

## Proposal Checklist

Before you start your work, you must submit a proposal for my approval. **Your proposal must include the following:**

_____ • the **question** you seek to answer

_____ • the **procedure** you will use

_____ • detailed **data tables** for recording measurements and calculations

_____ • a complete, itemized list of proposed **materials** and **costs** (including use of facilities, labor, and amounts needed)

**Proposal Approval:** ______________________________

(Supervisor's signature)

## Report Procedures

When you finish your analysis, prepare a report in the form of a business letter to Mr. Williams. **Your report must include the following:**

_____ • a paragraph describing the **procedure** you followed to test the DNA from each subject

_____ • a complete **data table** showing the $R_f$ values for each sample's RFLPs

_____ • a **graph** plotting the logs of the fragment lengths against $R_f$ values for the marker sample

_____ • your **conclusions** about whether the subjects may be related

_____ • a detailed **invoice** showing all materials, labor, and the total amount due

## Safety Precautions

- Wear safety goggles and a lab apron.
- Glassware is fragile. Notify your teacher promptly of any broken glass or cuts. Do not clean up broken glass or spills unless your teacher tells you to do so.
- Never use electrical equipment around water or with wet hands or clothing. Never use equipment with frayed cords.
- Wash your hands before leaving the laboratory.

## Disposal Methods

- Dispose of waste materials according to instructions from your teacher.
- Place used paper towels and plastic zippered bags in a trash can.
- Place broken glass, microtubes with DNA samples, gel, unused 1× TBE buffer, unused loading dye, unused DNA stain, and disinfectant solution in the separate containers provided.
- Wash reusable materials such as glassware and lab utensils, and return them to the supply area.

## FILE: Williams & Associates

**MATERIALS AND COSTS** (Select only what you will need. No refunds.)

### I. Facilities and Equipment Use

| Item | Rate | Number | Total |
| --- | --- | --- | --- |
| facilities | \$480.00/day | | |
| personal protective equipment | \$10.00/day | | |
| clock or watch with second hand | \$10.00/day | | |
| gel-casting tray (with gel in plastic bag) | \$5.00/day | | |
| 6-well gel comb | \$2.00/day | | |
| microtube rack | \$5.00/day | | |
| 10 μL fixed-volume micropipetter | \$2.00/day | | |
| gel spatula | \$3.00/day | | |
| staining tray | \$5.00/day | | |
| gel electrophoresis system | \$10.00/day | | |
| freezer | \$10.00/day | | |
| metric ruler | \$1.00/day | | |
| calculator | \$5.00/day | | |

### II. Labor and Consumables

| Item | Rate | Number | Total |
| --- | --- | --- | --- |
| labor | \$40.00/hour | | |
| graph paper | \$0.25/sheet | | |
| microtubes with cut DNA samples | \$10.00 each | | |
| 1–5 μL micropipets | \$2.00 each | | |
| 5–200 μL micropipetter tips | \$2.00 each | | |
| 1× TBE buffer | \$0.20/mL | | |
| loading dye | \$0.10/μL | | |
| DNA stain | \$0.10/mL | | |
| distilled water | \$0.10/mL | | |
| disinfectant solution | \$2.00/bottle | | |
| paper towels | \$0.10 each | | |
| nontoxic, permanent marker | \$2.00 each | | |
| **Fines** | | | |
| OSHA safety violation | \$2,000.00/incident | | |
| | | **Subtotal** | |
| | | **Profit Margin** | |
| | | **Total Amount Due** | |

Name ______________________________

Date ________________ Class ________________

HOLT BIOSOURCES
LAB PROGRAM
BIOTECHNOLOGY

# D9 Laboratory Techniques: Introduction to Fermentation Technology

## Skills

- using aseptic technique
- culturing bacteria

## Objectives

- *Assemble* a working fermenter.
- *Produce* the antibiotic bacitracin using *Bacillus licheniformis* in the fermenter.
- *Perform* a qualitative test of the antibiotic produced by *Bacillus licheniformis.*

## Materials

- safety goggles, gloves, and a lab apron
- petroleum jelly
- 3-hole rubber stopper, #8
- rigid plastic tubing, 11 in. length, 3/16 in. outside diameter (2)
- rigid plastic tubing, 5 in. length, 3/16 in. outside diameter
- vinyl air-line tubing, 3 in. length, 5/16 in. outside diameter (3)
- 1 L Erlenmeyer flask
- 70% ethanol or isopropanol (rubbing alcohol)
- in-line filters (2)
- vinyl air-line tubing, 3 ft length, 5/16 in. outside diameter
- screw compress clamp
- agar slant of *Bacillus licheniformis*
- sterile inoculating loops (2)
- 10 mL graduated cylinder
- 100 mL bottle of sterile NYSM inoculum medium with a magnetic stir bar
- magnetic stirrer
- 500 mL bottle of sterile NYSM fermentation medium
- air pump
- 10 mL syringe without a needle
- test tube, 13 × 100 mm
- sterile pipet
- pH paper or pH meter
- petri dish of *Bacillus thuringiensis* var. israelensis (2)
- forceps
- sterile 5 × 50 mm pieces of filter paper for antibiotic assay (12)

## Purpose

You have just been hired as an industrial microbiologist for a large pharmaceutical company. The company is a major supplier of the antibiotic bacitracin. Your job will be to monitor the production of bacitracin and to conduct research to improve the fermentation technology by which it is produced. To learn about the production of bacitracin through fermentation, first you will observe the process using a fermenter made from simple lab equipment. Then, you will assay, or test, the product you made for its ability to kill bacteria.

**PREPARATION NOTES**

**Time Required:** 2–3 50-minute periods; 15 minutes each on days 3, 4, and 5 for assays and data collection

**Materials**
Materials for this lab activity can be purchased from WARD'S. See the *Master Materials List* for ordering instructions.

**Preparation Tips**

- **Sterilization Procedure** (To be performed by the teacher ONLY.) Prior to culture inoculation and operation of the fermenters, the vessel (flask) and the rubber stopper must be sterilized either in a pressure cooker at 121°C (15 psi pressure) for approximately 15 minutes or by boiling them in a large pot for about 30 minutes. Wear a face shield and heat-protective gloves when handling hot objects.
- After heat sterilization, carefully cover the flask with a piece of aluminum foil. Allow it to cool completely before handling it.
- DO NOT heat sterilize the rigid plastic tubing; instead, prior to use, soak it in ethyl alcohol for about 5 minutes, as outlined in step 4 of the procedure.

## Background

- Provide no more than 200 mL of 70% alcohol in capped bottles.
- Purchase pH paper with a range of 6–9.
- Purchase prepared, sterile NYSM media and agar plates.
- To prepare NYSM plates with lawns of *B. thuringiensis,* follow the procedure below.

**1.** With a sterile inoculating loop, take a very small amount of the *B. thuringiensis* culture off the slant, and shake this into a test tube containing 9 mL of sterile saline.

**2.** Take several drops of this solution, and spread it with a bent stainless steel plating rod onto an NYSM agar plate.

### Additional Background

In 1945, the name *bacitracin* was given by Johnson, Anker, and Meleney to an antibiotic produced by an organism of the *Bacillus subtilis* group (later shown to be *Bacillus licheniformis*). This organism was isolated from tissue taken from a compound fracture of a tibia. The new antibiotic was water-soluble, had a low toxicity to animals, and resembled penicillin in its range of antibacterial activity. Bacitracin was originally discovered by B. A. Johnson in the bacteriological research laboratory of the Department of Surgery of Columbia University in June 1943. The first human case was treated on October 11, 1943, by the injection of the crude filtrate into the center of a furuncle of the nose "with prompt

The term *fermentation* has two meanings. In cell metabolism, the term *fermentation* refers to a metabolic process that follows glycolysis (the splitting of glucose molecules) in the absence of oxygen. Although fermentation occurs in all types of cells, the most familiar example of fermentation is the production of carbon dioxide and ethyl alcohol by microorganisms such as bacteria and fungi. The breakdown of organic compounds during cell metabolism also enables cells to obtain the energy and raw materials they need to make other organic compounds. The term *fermentation* is also used in the field of **biotechnology,** the application of biological techniques to industry. In biotechnology, **fermentation** is the controlled production of commercially useful products by the action of microorganisms. These products are produced in containers called **fermenters.** Products resulting from fermentation include beer, wine, cheese, citric acid, solvents, insecticides, fragrances, and vitamins, as well as a host of antibiotics, enzymes, and other organic compounds with substantial commercial value.

Recombinant DNA technology and the science of genetic engineering have increased the number of commercially important chemicals that can be produced by fermentation technology. The genes that code for the production of a number of substances have been identified and introduced into well-understood organisms such as *E. coli.* These organisms can be economically mass-produced to increase the yield of desirable products. Today, virtually all antibiotics are produced by isolating and purifying products made by organisms grown in fermenters. For example, bacitracin, which was discovered at almost the same time as penicillin, is produced through fermentation by the bacterium *Bacillus licheniformis.* **Bacitracin** is a polypeptide antibiotic that acts as a cell wall inhibitor, particularly of gram-positive bacteria.

### *Part 1—Fermentation Vessel Assembly*

1.  Put on safety goggles, gloves, and a lab apron.
2. Apply a small amount of petroleum jelly to each hole of a #8, 3-hole rubber stopper. Carefully insert three pieces of rigid plastic tubing (two 11 in. pieces and one 5 in. piece) through the holes of the rubber stopper, as shown in the diagram at right. The shorter rigid tubing will be the air outlet. One of the two longer tubes will be the air inlet, and the other will be a sampling port.

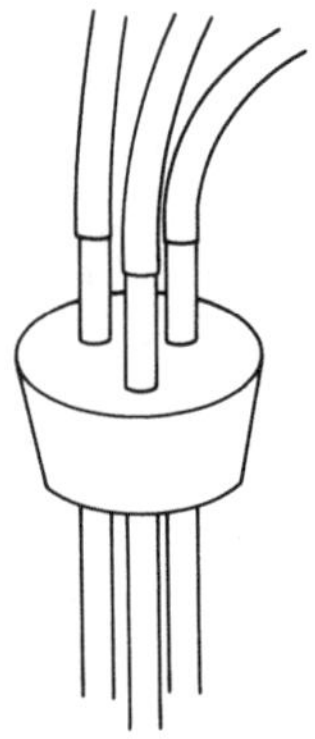

3. Attach a 3 in. piece of vinyl tubing to the top end of each piece of rigid tubing, as shown in the diagram above.
4.  Soak the top (stopper and tubing) assembly you made in steps 2 and 3 in 70% ethanol for about five minutes to sterilize it. Prior to use, remove the assembly from the alcohol, hold it over a sink, and quickly shake off any excess alcohol still present on the assembly and in the tubes. **CAUTION: Be careful not to get any alcohol on you or your classmates.**

resolution of the infection." Unfortunately, bacitracin was later found to be too toxic. Since then, it has been successfully used in combination with neomycin and other antibiotics for topical use in cuts, burns, and other surface wounds.

What is particularly interesting about the biosynthesis of these compounds is that they involve a novel assembly mechanism in which the amino acid sequence is determined and the peptide bond is formed by specific enzymes, and neither tRNAs (transfer RNA) nor ribosomes participate. The role of these "antibiotics" in the producer is not known, but it is believed that they may control various stages of the spore differentiation process during sporulation.

In the initial part of the pH-curve for the *Bacillus* life cycle, there is a drop in pH (culture becomes acidic). This is due to the use of sugars and the subsequent formation and release of acid products formed during cell growth and division. As the carbon source (sugars and amino acids) found in the medium is used up, sporulation generally commences, and the acidic products formed during vegetative growth (the period before spores are formed) are used in the spore formation. We then see a rise in the pH curve (it becomes more alkaline). This tells the fermentation scientist when sporulation has begun and whether the fermentation is following its prescribed path.

5. Immediately place the top assembly onto a flask, making sure that the rubber stopper seals the opening of the flask completely.

6. Attach an in-line filter to each end of the air-inlet and air-outlet tubings, and then attach a 3 ft piece of vinyl air-line tubing to the filter on the air-inlet tubing. Finally, attach a screw compress clamp to the sampling port, as shown in the diagram below. Your fermenter is now ready for use.

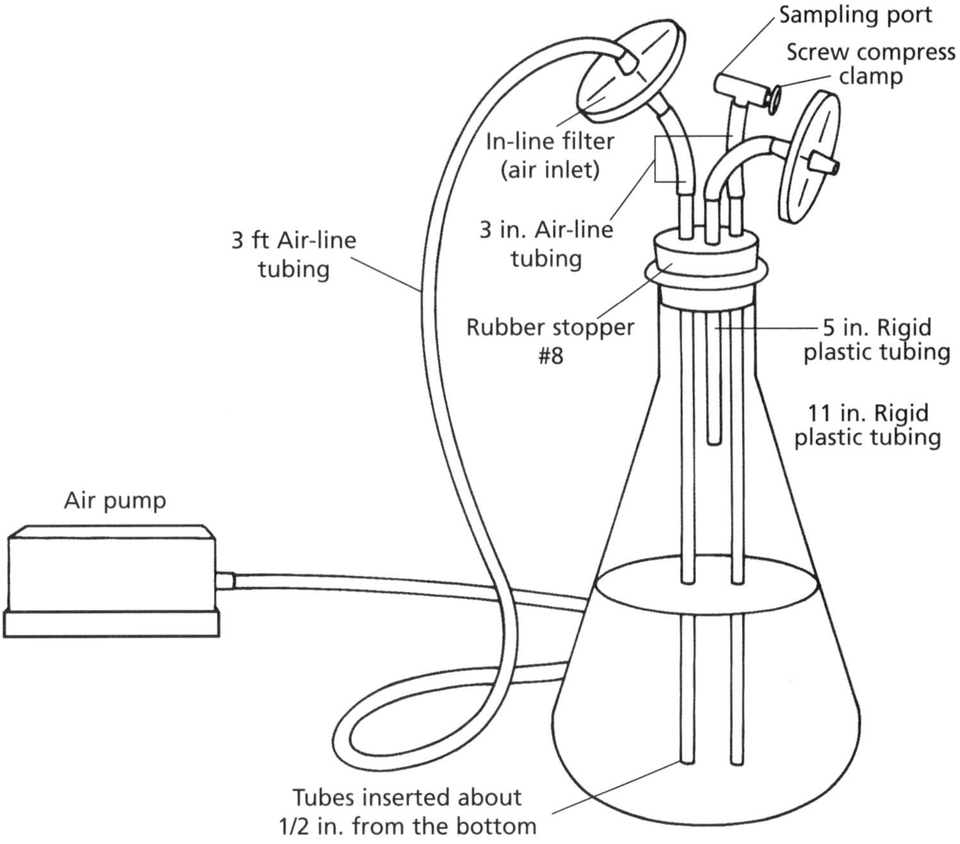


## *Part 2—Preparing Inoculum and Inoculating the Fermenter*

Before beginning the process of fermentation, an inoculum, which is a quantity of bacteria, must be "built up." The inoculum buildup has the following three purposes: to maximize the number of cells; to synchronize their growth; and to prepare cells that are in the proper growth stage. The life cycle of a colony of bacteria can be divided into the following three stages: vegetative growth, sporulation (spore formation), and death. Some of the useful products bacteria produce through fermentation are made during the vegetative stage, while others, such as bacitracin, are made during the sporulation stage.

7. 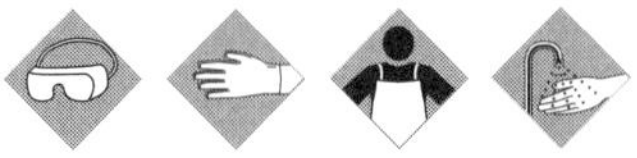 Put on safety goggles, gloves, and a lab apron. *Note: Although these microbes are naturally occurring and nonpathogenic, you must use personal protective equipment and aseptic technique throughout the procedure to minimize the risk of any contamination.*

8. Measure 1 mL of sterile NYSM inoculum medium, and add it to an agar slant of *B. licheniformis.*

### Disposal

- All leftover cultures, the fermentation broth, and any materials that have come in contact with bacteria must be decontaminated prior to disposal. Instruct students to place all used (contaminated) materials into a biohazard bag for disposal. Decontamination may be accomplished by autoclaving at 121°C and 15 psi for 15 minutes.
- If an autoclave is not available, you may treat all contaminated materials and leftover cultures by soaking them in full-strength bleach solution. Allow them to soak for 30 to 60 minutes prior to disposal.

### PROCEDURAL NOTES

### Safety

- Discuss all safety symbols and caution statements with students.
- Although these microbes are naturally occurring and nonpathogenic, it is recommended that you have students use safety goggles and gloves throughout the procedures to minimize the risk of contamination.
- The cultures provided in this experiment are not reported to be pathogenic to humans; in fact, they have been specifically selected as the most innocuous materials available that will still suit the aims of these exercises. In any given population, however, there may be an individual who is allergic to the organisms, the media, or one of the reagents. Carefully monitor students, and remind them to use aseptic technique at all times during this activity.

9. Using a sterile inoculating loop, carefully scrape off the growth from the slant into the liquid. Pour the liquid containing the bacteria back into the 100 mL NYSM inoculum bottle.

10. Make sure the cap of the 100 mL NYSM inoculum bottle is screwed on tightly, and then place the bottle on its side on a magnetic stirrer.

11. Incubate the inoculum bottle at room temperature or in a 30°C incubator, if available. When growth becomes apparent (usually within 18–24 hours), the inoculum is ready and can be used to inoculate (add bacteria to) your fermenter.

12. Working quickly, partly remove the top assembly from the flask of your fermenter by pulling up on one of the rigid tubings with one hand. With your other hand, pour the contents of a 500 mL bottle of NYSM fermentation medium into the flask.

13. Attach the air-inlet tubing to an air pump, and adjust the air flow so that air bubbles gently into the medium. This will provide oxygen to cells as well as provide a positive pressure to prevent contamination.

14. Using the technique described in step 12, carefully add the inoculum you made in steps 8–11 to your fermenter. Allow the fermentation process to run for 48 hours.

## *Part 3—Sampling and Testing Procedures*

To monitor the progress of your fermentation run, you will need to periodically withdraw small samples of your fermentation broth (inoculated fermentation medium) and test the samples. Test the pH of your broth at the start of the fermentation run, at least once during the vegetative phase (at about 24 hours), and again when the broth is harvested (at about 48 hours). Assay (test) for bacitracin production after 24 hours and again at the end of 48 hours.

Almost all *Bacillus* bacteria, when grown in a liquid medium, show a distinctive change in pH during their growth cycle. An example of this change is shown in the graph below.

**Growth and pH Curves for *Bacillus* in a Fermenter**

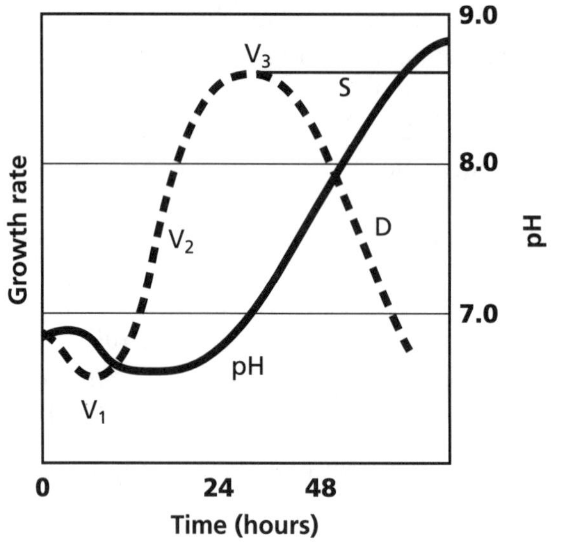


**Phases of Growth**
V = Vegetative growth
$V_1$ = Phase of adjustment
$V_2$ = Phase of exponential growth
$V_3$ = Phase of no growth
S = Sporulation
D = Death (phase of decline)

**Procedural Tips**

- Be sure that the fermentation medium students add to their fermenters is at room temperature.
- One way to increase the rate of growth is to increase the amount of airflow in the fermenter (and vice versa). Do not allow foaming to occur.
- There should be an emphasis on good laboratory practice and laboratory safety, less so because of any risk to the student than because of the need to ensure that the experimental materials do not become contaminated. Working carefully rather than quickly is the key to success.
- The experiment itself may be divided into more than one part, with each part being done by a different group. In fact, this may be advisable for the sampling necessary for various activities, such as determining pH, spore staining, and bioassays, that require one or more time periods. Although we strongly urge you to follow the experiment as written, there may be room for some improvisation. For example, extra sampling times might be added if materials permit and time warrants.
- If you want students to use a pH meter instead of pH paper to test for pH, have them follow this procedure:
  1. Place 1 mL of culture material in a small beaker.
  2. Add enough distilled water to allow the pH probe to be fully immersed (5–10 mL).
  3. Record the pH value in the data table.
  4. Completely rinse off the pH probe at the end of each test.

**15.** To remove a sample of fermentation broth, attach a syringe (without a needle) to the sampling port of your fermenter. Loosen the screw compress clamp, and pull on the syringe handle to withdraw 1–2 mL of broth.

**16.** Expel the sample into a small test tube, and label it accordingly.

**17.** Tighten the screw compress clamp, and rinse the end of the sampling port with a small amount of either 70% ethanol or rubbing alcohol to prevent the contamination of your fermenter.

**18.** Using a sterile pipet, place a drop of the broth culture on a piece of pH paper. Record the pH value in the data table below.

**19.** Repeat steps 15–18 at the time periods shown in the table below.

***Results of pH Test and Bacitracin Assay***

| Variable | Incubation time | | |
|---|---|---|---|
| | Start | 24 hours | 48 hours |
| pH | 6.7 | 7.0–7.6 | 8.0–8.6 |
| Bacitracin assay (size of zone of inhibition) (mm) | —— | some growth, with slight zone of inhibition around test strip with bacitracin | moderate growth, with well-defined zone of inhibition around test strip with bacitracin |

**20.** Use the remainder of the samples removed after 24 and 48 hours to assay for bacitracin. Using forceps, dip a sterile strip of filter paper into the sample. Then carefully place the strip onto the agar in the center of a petri dish previously inoculated with *B. thuringiensis,* as shown in the diagram below.

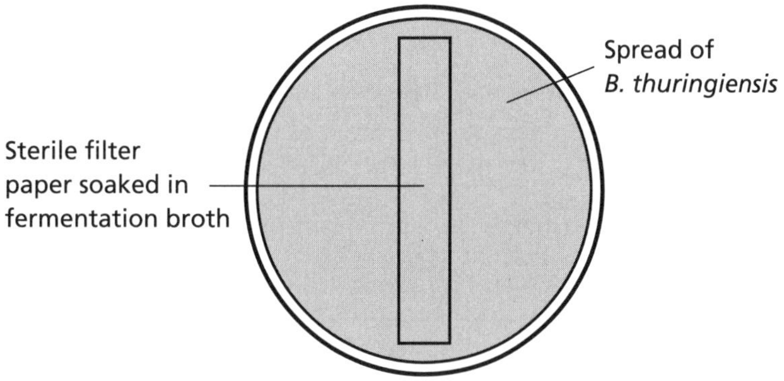


**21.** Incubate each petri dish until growth is visible (usually within 48 hours). At that point, you should see a zone of inhibition (a clear zone of no growth) if bacitracin is present.

**22.** Dispose of your materials according to the directions from your teacher.

**23.** Clean up your work area and wash your hands before leaving the lab.

## Analysis

• Students should see little or no inhibition on the NYSM plate during the vegetative-growth phase, but they should see increasing amounts of bacitracin activity during the sporulation phase.

• As a control for the bacitracin assay, you may want students to place a 0.1 mL drop from the fermenter broth onto an uninoculated NYSM plate and incubate the plate for 48 hours.

**24.** What was the purpose of the in-line filters in your fermenter?

The filters remove any possible contaminants that may enter or leave the fermenter.

**25.** What happened to the pH of the fermentation broth during the fermentation run? How do you account for this change?

The pH of the broth increased (became less acidic) as amino acids were removed from the solution and were used to produce bacitracin, which is a polypeptide (chain of amino acids).

## Conclusions

Like most biological experiments, many things could go wrong. The following are potential problem areas:

a. The larger the bacterial population is, the greater the amount of antibiotic must be to inhibit the growth of the bacterial lawn.

b. If growth in the fermenter is heavy, there may be outgrowth from the antibiotic strip that has been applied to the NYSM plate. One should still see a zone of inhibition around this growth.

**26.** Do you think that the production of antibiotics is important to the survival of bacteria in nature? Justify your answer.

Answers will vary. Antibiotics are probably not important to the survival of bacteria in nature, except in rare cases. A single bacterium would not produce much antibiotic. It is mainly when humans produce a large number of bacteria that are all in the same growth phase that a high concentration of antibiotic toxic to other organisms can be produced.

**27.** Bacteria are used to produce compounds that are found only in certain species, such as human insulin. How do you think this is accomplished?

Answers will vary. Students should reason that recombinant DNA techniques could be used to isolate a human gene or genes responsible for the production of the product and insert the gene into a strain of bacteria. The bacteria could then be grown in a fermenter to produce large amounts of the product.

## Extensions

**28.** Search an on-line data base or do library research to learn about how recombinant DNA and industrial fermentation techniques are used in the production of other economically important chemicals.

**29.** *Microbiologists* study microorganisms, such as bacteria and protists. Find out about the training and skills required to become a microbiologist, and determine what types of industries employ microbiologists.

Name ______________________________

Date ______________ Class ______________

HOLT BIOSOURCES
LAB PROGRAM
BIOTECHNOLOGY

# D10 *Laboratory Techniques: Ice-Nucleating Bacteria*

## Skills

- making observations
- graphing data

## Objectives

- *Observe* the effects of ice-nucleating proteins on ice formation, the freezing temperature of water, and the heat of crystallization.
- *Calculate* the cooling rate of water.

## Materials

**PREPARATION NOTES**

**Time Required:** two 50-minute periods

- safety goggles
- lab apron
- wax pencil or felt-tip marking pen
- test tubes (3)
- test-tube rack
- distilled water (15 mL)
- ice-nucleating protein granules
- rubber stopper
- microplate
- aluminum foil (4 in. × 6 in.)
- graduated plastic pipets (4)
- freezer
- stopwatch or clock with second hand
- cardboard strips, 1/2 in. × 2 in. (2)
- −30°C to 50°C thermometers (2)
- stapler
- calculator

## Purpose

**Materials**

Materials for this lab activity can be purchased from WARD'S. See the *Master Materials List* for ordering instructions.

You are a plant physiologist working for the state department of agriculture. You live in a state that grows a majority of the country's oranges and lemons. Once oranges and lemons form on the trees, they are very susceptible to damage by freezing temperatures. One way growers protect their fruit from freezing temperatures is to spray the trees with water. The resulting layer of ice protects the fruit from further damage. You have read about a kind of bacteria, called ice-plus bacteria, that causes water to freeze at a higher temperature. You wonder if the oranges and lemons would have less damage if the protective coating of ice formed at a higher temperature. You decide to run a series of tests to find out at what temperature the ice-nucleating protein produced by ice-plus bacteria causes water to freeze.

## Background

**Additional Background**

The molecules in water are in continuous motion. The energy of this motion determines the temperature of the water and prevents any intermolecular structure (such as ice crystals) from forming. For freezing to begin, enough energy must be removed from the water to allow the molecules to slow

A great deal of research has been done lately concerning the freezing temperature of water. The common misconception is that water freezes at 0°C. In fact, freezing rarely starts at 0°C. Water in its purest state can be "supercooled" to as low as −40°C without ice formation.

An "ice nucleator" helps water to freeze by attracting the water molecules and slowing them down. A **nucleator** is any foreign particle in water that allows the freezing process to begin. The ice-nucleating protein (INP) that will be used in this investigation is derived from the naturally occurring bacterium *Pseudomonas syringae.* This form of *Pseudomonas* is sometimes called the "ice-plus" variety because it contains a gene that promotes the formation of proteins that serve as nucleators.

Another feature of an ice-nucleating protein is that it is capable of initiating the freezing process at a higher temperature. The result is that water treated with an ice-nucleating protein freezes faster, more completely, and over a wider range of conditions.

The ability to have ice form at higher temperatures has many commercial applications, such as weather modification (including snow making), natural cooling of water as a refrigeration source, water purification, and construction in the Arctic and Antarctic. On the other hand, the formation of ice causes wide-scale damage to some crops due to the presence of ice-plus bacteria.

However, another strain of *Pseudomonas,* an "ice-minus" strain, also occurs naturally. It is identical to the ice-plus variety except that it lacks the ice-promoting gene. Ice-minus bacteria do not freeze until the temperature goes below −7°C. Since the formation of ice causes frost damage, the absence of the ice-nucleating protein protects some plants from frost as much as seven degrees below the normal freezing point of water. When these ice-minus bacteria are applied to a plant, water can be supercooled rather than freeze and damage the plant. The economic incentive of protecting plants and crops from frost damage is enormous.

### *Part 1—Observing the Effects of Ice-Nucleating Protein*

1. Put on safety goggles and a lab apron.

2. Use a wax pencil or felt-tip marking pen to label a test tube "INP-treated water." **CAUTION: Glassware is fragile. Notify your teacher promptly of any broken glass or cuts. Do not clean up broken glass or spills unless your teacher tells you to do so.** Add 10 mL of distilled water to the test tube. Then add 3 to 4 granules of the ice-nucleating protein to the test tube. Put a rubber stopper in the mouth of the test tube. Mix the contents well by inverting the test tube several times.

3. With the wax pencil, write your name on the bottom of a microplate. Wrap the top of the microplate with the piece of aluminum foil. Make an imprint of the wells on the foil by gently rubbing the foil with your hand.

4. Using a graduated plastic pipet, place one drop of the INP-treated water into 30 of the wells formed by the foil. *Note: For the best results, keep the water droplets as small as possible.* Save the remaining INP-treated water for later use.

5. With a clean, graduated plastic pipet, place one small drop of distilled water into 30 adjacent wells formed by the foil. *Note: Again, keep the water droplets as small as possible.*

6. Place your microplate in a freezer. Make observations of your plate every three minutes for about 15 minutes or until one set of droplets freezes. Watch for any changes in freezing between the plain distilled water droplets and the INP-treated water droplets. *Note: Keep the freezer door open only long enough to make your observations.*

down and align in a pattern that allows ice crystals to form.

When ice-nucleating protein is mixed into a water supply at the right concentration, many water molecules are effectively "seeded" with a nucleation site. Water then attaches to the protein in an arrangement that mimics the structure of an ice crystal. Once a crystal forms, others will grow from this nucleus. The addition of this ice-nucleating protein to source water increases the number of nucleation sites by as much as 100,000 times. This means that every droplet of water has a site for ice crystals to grow.

## Procedure

Although both ice-plus and ice-minus *Pseudomonas* occur naturally, the ice-plus strain is dominant. Using genetic engineering techniques, scientists are able to remove the ice-promoting gene from the ice-plus strain and produce large numbers of the newly created ice-minus bacteria. In California in 1987, newly created ice-minus bacteria were sprayed on strawberry and potato fields in an attempt to make the ice-minus strain dominant over the ice-plus variety and thus reduce frost damage.

**Preparation Tip**

Prepare separate containers for the disposal of broken glass and liquids.

**Disposal**

Dilute solutions containing ice-nucleating protein in a ratio of 1 part solution to

20 parts water. Pour the diluted solution down the drain.

**PROCEDURAL NOTES**

**Safety Precaution**

Discuss all safety symbols and caution statements with students.

**Procedural Tips**

- To reduce the amount of time required to complete this investigation, students may start with chilled distilled water of about 10°C (50°F). Treat the water with INP just prior to use. The temperature of both tubes at the beginning of the investigation ($t_1$) should be the same.
- Determination of the rate of cooling and heat of crystallization may be postponed to another day if time does not allow for completion of the investigation in one laboratory period.
- The INP-treated water should start to freeze sooner and at a higher temperature than the plain distilled water.
- The heat of crystallization temperature change is approximately $^{+}3$°C for the distilled water and practically undetectable with less-precise thermometers for the INP-treated water.
- Although temperature readings may be different from student to student, they should approximate those in the data tables following step 12.

◆ Did the INP-treated water droplets freeze faster than the plain water droplets? Explain your answer.

INP-treated water should freeze faster than plain water because of the ability of the ice-nucleating protein to start ice crystal formation at a higher temperature than otherwise possible.

## *Part 2—Measuring the Effects of INP on Freezing Temperature, Cooling Rate, and Heat of Crystallization*

7.  Use a wax pencil to label two test tubes *Tube 1* and *Tube 2.* Add your initials to each tube.

8. Use a clean, graduated plastic pipet to transfer 3 mL of the INP-treated water to Tube 1. Place the tube in a test-tube rack.

9. Using a clean, graduated plastic pipet, add 3 mL of distilled water to Tube 2, and place the tube in the rack next to Tube 1.

10. Fold two cardboard strips in half lengthwise. Place the fold of one piece of cardboard over the top of one of the thermometers. Staple each end to hold the cardboard securely to the thermometer as shown in the diagram below. *Note: Be careful not to break the thermometer with the stapler.* Repeat for the second thermometer.

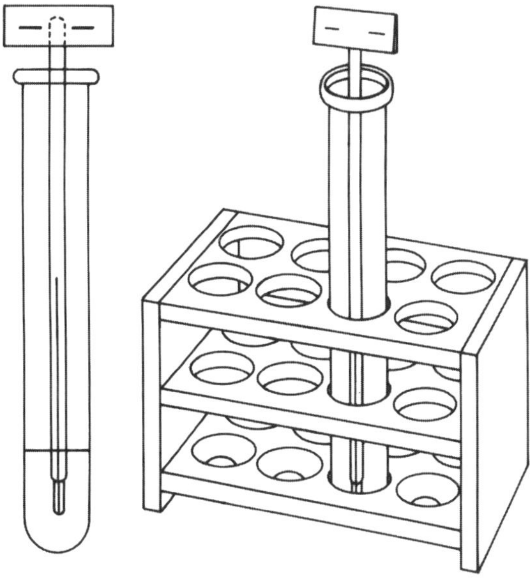

11. Insert a thermometer into each test tube. Use the cardboard strips to position each thermometer so that it does not touch the glass. Put the test tubes and test-tube rack in the freezer.

12. Take temperature readings of Tube 1 and Tube 2 at five-minute intervals for a minimum of 50 minutes. Do not touch the thermometers or tubes. Make careful observations of any ice formation during this period. *Note: Keep the freezer door open only long enough to take measurements.* Record the temperature readings and observations in the tables on the next page.

• Students who elect to do Extension step 24 should include information on animals surviving freezing by controlling ice formation. Ice formation must be initiated in extracellular fluids in a way that keeps the rate of freezing low and the size of the crystals small. Ice-nucleating proteins synthesized during autumn months provide the binding sites that order water molecules into an ice-lattice structure. In so doing, they keep the rate of freezing slow and the size of the crystals small. When freezing is a relatively slow and controlled event, it allows extra time for cells to adjust both physically and metabolically during the transition to the frozen state.

### *Tube 1 INP-Treated Water*

| Time (min.) | Temp. Tube 1 (°C) | Cooling Rate (°C/min) | Observations |
|---|---|---|---|
| 0 | 18 | NA | |
| 5 | 8 | −2.0 | |
| 10 | 1 | −1.4 | ice starting to form |
| 15 | 0 | −0.2 | |
| 20 | 0 | 0 | |
| 25 | 0 | 0 | |
| 30 | 0 | 0 | |
| 35 | 0 | 0 | |
| 40 | 0 | 0 | solid ice |
| 45 | −1 | −0.2 | |
| 50 | −1 | 0 | |
| 55 | −2 | −0.2 | |
| 60 | −2 | 0 | |

### *Tube 2 Distilled Water*

| Time (min.) | Temp. Tube 2 (°C) | Cooling Rate (°C/min) | Observations |
|---|---|---|---|
| 0 | 18 | NA | |
| 5 | 7 | −2.2 | |
| 10 | 2 | −1.0 | |
| 15 | 0 | −0.4 | |
| 20 | −1 | −0.2 | |
| 25 | −3 | −0.4 | ice starting to form |
| 30 | −2 | −0.2 | |
| 35 | 0 | +0.4 | |
| 40 | −1 | −0.2 | |
| 45 | −1 | 0 | |
| 50 | −2 | −0.2 | |
| 55 | −2 | −0.2 | solid ice |
| 60 | −3 | −0.2 | |

**13.** Calculate the rate of cooling for each test tube using the following equation. Record your findings in the appropriate data table on the previous page.

$$\frac{°C}{min} = \frac{T_2 - T_1}{t_2 - t_1}$$

where $T_1$ = temperature at time interval *start* $t_1$ = time interval *start*
$T_2$ = temperature at time interval *stop* $t_2$ = time interval *stop*

**14.** On the grid below, graph temperature versus time during freezing for Tube 1 and Tube 2. Using the graph, determine the heat of crystallization. The heat of crystallization is the amount of heat released when a liquid is transformed into ice crystals. The heat of crystallization for this lab is found by finding the lowest temperature at which water starts to freeze and the highest temperature at which freezing is completed. Find the difference between the two temperatures. Multiply that number by 18 to determine the heat of crystallization.

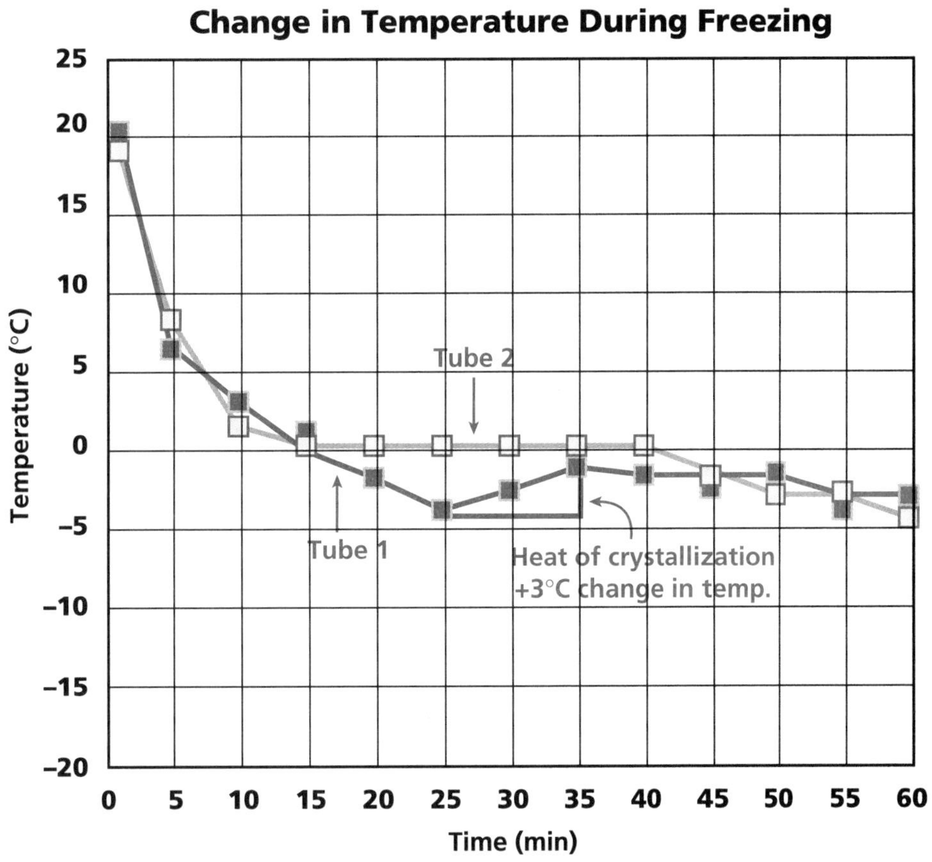


**15.** Dispose of your materials according to the directions from your teacher.

**16.** Clean up your work area and wash your hands before leaving the lab.

## Analysis

**17.** Which of the two test tubes from step 11 first started to display signs of ice crystal formation?

Test tube 1 containing the INP-treated water should display the first signs of ice formation.

**18.** At what temperature did the distilled water and the INP-treated water begin to freeze? Were these temperatures what you expected them to be?

The first signs of ice formation of the plain distilled water were evident at about −3°C, compared with the INP-treated water, which started to display signs of freezing at about 1°C. No, most would expect water to begin freezing at 0°C.

**19.** Explain the differences in the temperatures for the first signs of freezing.

For freezing to be initiated, there must be a nucleation site for ice crystals to grow. INP provides the initial nucleation site for ice crystal formation. Once ice formation has been initiated, ice itself can serve as a nucleator, and thus INP would not be needed after the initial ice formation stage. Distilled water requires a lower temperature than INP-treated water to start freezing because it lacks the nucleation sites provided by the INP molecules.

## Conclusions

**20.** According to your data and graph, what effect did the INP have on the formation of ice? the cooling rate of water?

The temperature at which ice formed in the INP-treated water was higher (1°C) than that of distilled water (−3°C). INP slows the cooling rate of water; the cooling rate in Tube 1 was greater (−0.33°C/min) than that of the INP-treated water in Tube 2 (−0.35°C/min).

**21.** What other substances, organic or inorganic, might be used as ice nucleators?

Answers will vary but might include microscopic particles such as dust, smoke, and other particles in the atmosphere that can act as ice nulceators by allowing water droplets to form around them. Chemicals such as silver iodide have also been used to seed clouds, causing precipitation of rain.

**22.** Based on the results of your experiments, what would be the next step in determining if oranges and lemons would be protected by the addition of ice-plus bacteria?

Answers will vary but should mention that the next step would be to spray some trees with INP-treated water and other trees with regular, untreated water and to compare the quality of the fruit following ice formation.

## Extensions

**23.** Find out how the application of ice nucleators can be used to modify weather, specifically how they can be used to cause rain or snow to fall.

**24.** Find out how ice-nucleating proteins are used in nature by animals such as insects, amphibians, and reptiles to help them survive freezing during winter.

Name ______________________________

Date ______________ Class ______________

HOLT BIOSOURCES
LAB PROGRAM
BIOTECHNOLOGY

# D11 Laboratory Techniques: Oil-Degrading Microbes

## Skills

- practicing aseptic technique
- developing a dynamic model of an oil spill
- controlling variables in an experiment
- collecting, organizing, and graphing data

## Objectives

- *Compare* the physical characteristics of oil before and after the action of oil-degrading microbes.
- *Identify* which microorganisms are useful in cleaning up an oil spill.

## Materials

- safety goggles
- gloves
- lab apron
- disinfectant solution in squeeze bottle
- paper towels
- plastic jars with lids (3)
- wax pencil
- distilled water (90 mL)
- refined oil in dropper bottle
- 1.5 g of nutrient fertilizer
- scoop
- disposable pipets (2)
- *Pseudomonas* culture
- *Penicillium* culture
- density indicator strips (3)
- incubator
- biohazard waste disposal container

**PREPARATION NOTES**

**Time Required:** one 50-minute period to set up the experiment and 10–15 minutes over the next 4–7 days to collect data. *Note: Cultures grown at 30°C will achieve faster results than those grown at a cooler temperature. The length of the experiment may need to be adjusted for this variable.*

## Purpose

You are a marine ecologist who works for a large petroleum company. One of the problems you must solve is how to clean up the environment following an oil spill. You believe that some of the present, mechanical methods of cleaning up oil are not effective. You want to explore new methods. You have heard that certain kinds of microbes can be used to digest the spilled oil. Before an oil spill happens, you want to find out if microbes really can digest oil. You set up a controlled experiment to find out which of two different microbes digests oil most efficiently.

**Materials**

Materials for this lab activity can be purchased from WARD'S. See the *Master Materials List* for ordering instructions.

## Background

Through the media, the public has been made increasingly aware of the hazards to the environment caused by oil spills. A single gallon of oil can spread thinly enough to cover four acres of water. Some of the oil evaporates; some is broken down by radiant energy; and some emulsifies, or breaks down into small pieces, to form a heavy material that eventually sinks to the bottom of the ocean. This heavy material endangers birds, marine mammals, and other forms of sea life.

Oil spills can be cleaned up using mechanical devices such as skimmers and barriers. Other cleanup methods include the use of chemical dispersants and solvents and the burning of oil. However, some microorganisms can break down the various hydrocarbons in an oil spill. They may be the best, most environmentally safe prospect for cleaning up oil spills. These "oil hungry" microbes convert oil into food for themselves, rendering it nontoxic and allowing it to be assimilated safely into aquatic food webs. The use of living organisms to repair environmental damage is known as **bioremediation.**

**Additional Background**

When an oil spill occurs, oil floats on the water's surface and is moved from the spot of the spill by wind and waves. Approximately 25 percent of the spill is lost almost immediately to evaporation. The remaining oil becomes very thick and sinks to the ocean floor, where it is acted on by microorganisms and the sun. Within a few

## Procedure

months, only about 15 percent of the oil remains. This oil, however, is the material that adheres to objects such as rocks and sand. If the spill occurs near land, this thick material adheres to the beach, rock, birds, and other marine life in the region.

An oil spill has an immediate impact on the food chain by reducing the amount of light than can penetrate the upper layer of the water. Studies indicate that 2 m below the surface, light is reduced by about 90 percent. Photosynthesis takes place within the upper 27 m of the ocean's surface. Other effects include a reduction in the amount of dissolved oxygen, damage to marine birds, and damage to the marine environment due to the toxic compounds contained in the oil.

The refined oil used in this investigation is light brown in color and forms a smooth and continuous layer on the water's surface.

**Preparation Tip**

Rehydrate the growth culture 4–5 days prior to use in the activity. You may also have your students participate in the procedures of rehydration and growing of cultures. Rehydration instructions come with the special oil-degrading, freeze-dried *Pseudomonas* and *Penicillium* cultures.

### *Part 1—Inoculating Oil With Microorganisms*

1. Put on safety goggles, gloves, and a lab apron.

2. Use disinfectant solution to sterilize the top of your lab table. To do this, spread disinfectant solution over the entire work area. Wipe the area clean with paper towels. Dispose of the paper towels as indicated by your teacher.

   ♦ Why is it important to sterilize your work area before you begin this investigation?

   To prevent contamination of the test cultures by microorganisms that might be present on the lab table.

3. Use a wax pencil to label one plastic jar *Control*, a second plastic jar *Pseudomonas*, and a third plastic jar *Penicillium*. Also write the date and your name and class period on each jar.

4. Pour about 30 mL of distilled water into each jar so that each jar is about half full.

5. Add about 20 drops of refined oil to form a thin layer in each jar.

6. Using a scoop, add a pinch of fertilizer to the water and oil mixture. This fertilizer is the kind used in real oil spills and will coagulate the oil and provide nutrients for microbial growth.

7. Using disposable pipets, inoculate the appropriate jars with 5 mL (two droppers full) of a *Pseudomonas* culture and 5 mL of a *Penicillium* culture. The control jar receives no microorganisms. **CAUTION: Always practice aseptic technique when using bacteria in the lab.**

8. Secure the top on each jar, and invert it several times to mix the oil with the jar's other contents. Similar wave action occurs in the ocean and increases the amount of dissolved oxygen in the water.

   ♦ Why is it desirable to increase the amount of dissolved oxygen in the water?

   The microorganisms will grow more readily in well-oxygenated water.

9. Number the bars on three density indicator strips from 1 to 5, with 5 for the darkest. Place a density strip on each jar so that the tops of the bars are below the water level. This density strip will allow you to quantify the amount of microbial growth in each jar. As the microbes grow, the water will become cloudy, or turbid. As the microbe population increases, you will be able to see fewer bars through the water. Hold up the jar with the density strip opposite you. Place a piece of white paper behind the jar to provide a background. Look through the water in the jar to determine the number of density bars that have disappeared.

## Disposal

- When students have completed this lab, autoclavable materials should be sterilized in either an autoclave or a pressure cooker. All materials in contact with bacteria should be autoclaved at 121°C and 15 psi for 20 minutes before disposal.
- Solid trash that has been contaminated by biological waste must be collected in a separate and specially marked disposal bag. Package all sharp instruments in separate metal containers for disposal. No pipets should protrude through the disposal bag. After autoclaving, disposal bags should be placed in an outer sealed container (plastic bucket with lid).

## PROCEDURAL NOTES

### Safety Procedures

- Discuss all safety symbols and caution statements with students.
- Tell students that although these oil-degrading microbes are naturally occurring and nonpathogenic, they must wear safety goggles, a lab apron, and gloves throughout the procedure to minimize the risk of contamination.
- Review the safety precautions for handling microorganisms in the lab. Remind students to use aseptic technique at all times during this lab.
- Make sure students wash their hands with antibacterial soap before and after the lab and sterilize their work areas before and after the investigation.

**10.** Record your observations for today, day 0, in the data table below.

***Effects of Microorganisms on Oil***

| Day | Organism | General appearance of oil | Color of oil | Turbidity of water (number of bars that disappear) |
|---|---|---|---|---|
| 0 | *Pseudomonas* | large blob in the middle of the jar | light brown | 1 |
| | *Penicillium* | large blob in the middle of the jar | light brown | 1 |
| | Control | large blob in the middle of the jar | light brown | 1 |
| 1 | *Pseudomonas* | large blob in the middle of the jar | not as dark as day 0 | 1 |
| | *Penicillium* | large blob in the middle of the jar | not as dark as day 0 | 2 |
| | Control | large blob in the middle of the jar | light brown | 1 |
| 2 | *Pseudomonas* | smaller blobs of oil | yellowish brown | 2 |
| | *Penicillium* | smaller blobs of oil | yellowish brown | 3 |
| | Control | large blob in the middle of the jar | light brown | 1 |
| 3 | *Pseudomonas* | oil blobs do not appear as thick | ivory | 2 |
| | *Penicillium* | oil blobs do not appear as thick | ivory | 3 |
| | Control | large blob in the middle of the jar | light brown | 1 |
| 4 | *Pseudomonas* | thin drops | ivory | 3 |
| | *Penicillium* | thin drops | ivory | 4 |
| | Control | large blob in the middle of the jar | light brown | 1 |

**11.** Incubate the jars with caps half loosened at 30°C. If an incubator is not available, place the jars in a warm spot in the classroom.

◆ Why is it necessary to incubate the jars?

Incubation is necessary to allow the organisms time to multiply. A well-developed culture will degrade the oil more effectively.

**Procedural Tips**

- Divide the class into teams of two to four students.
- Set up a culture-growing station with a plastic pipet at each culture for students to use as they inoculate their test tubes.
- Although specific observations may vary from student to student, they all should agree that the oil starts showing signs of degradation after 1–2 days of microbial action, based on its color change from deep brown to yellowish brown, as well as changes in its physical characteristics. The turbidity of the solutions should increase, indicating microbial growth. In this case, the fertilizer adds an additional carbon source (nutrients) for the microbes.

◆ Why is it necessary to leave the jar caps half loosened during incubation?

The jar caps must remain half loosened to allow movement of oxygen and carbon dioxide in and out of the jars.

12.  Dispose of your materials according to the directions from your teacher.

13.  Sterilize your work area as described in step 2. Wash your hands with antibacterial soap before leaving the lab.

## *Part 2—Observing the Effects of Microbes on Oil*

14. Observe your tubes every 24 hours for 4 days. Look for any signs of oil degradation, such as change in color, the formation of tiny oil droplets, break-up of the oil layer into smaller fragments, or changes in texture. Microbial growth will be observed mainly at the interface (boundary) between the oil and water. Record your observations of the general appearance of the degrading oil, its color, and the turbidity of the water in the data table on the previous page. Each day after you make your observations, temporarily tighten each jar lid and invert each jar once or twice to increase the dissolved oxygen content in the water and to further mix the oil with the jar's other ingredients.

◆ Which microorganisms do you think will degrade the most oil?

Answers will vary.

## Analysis

15. Using the grid below, prepare a graph showing the change in turbidity that occurs in each jar. The turbidity (number of bars that disappear) should be shown on the *y*-axis, with time plotted on the *x*-axis. Use different colored pencils to draw the curve for growth of both organisms and the control on the same graph. Be sure to make a legend that indicates which color pencil represents which organism and the control.

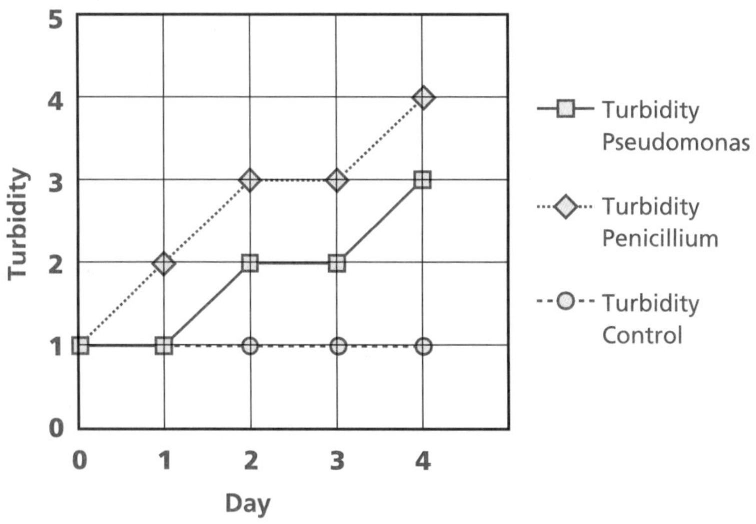

16. Describe any changes in the physical characteristics and appearance of the oil on day 1 and beyond. Discuss the possible causes for such changes.

The oil should start to show signs of degradation after 1 or 2 days of microbial action, changing color from dark brown to yellowish brown and eventually to ivory. Another change in its physical characteristics is the breakdown of the continuous layer of oil on the water's surface into minute oil droplets. The control jar should show no change.

17. What does an increase in turbidity indicate?

An increase in turbidity indicates an increase in microbial population as well as degradation of the oil.

18. What is the purpose of the control?

The control shows that the change in the microbe-containing oil—both in color and in texture—was due to the microbes. While the control showed no change, the color of the microbe-containing oil became lighter, and the oil broke down into smaller pieces.

19. What is the purpose of starting with distilled water rather than tap or pond water?

Tap water may have chlorine in it that may kill the bacteria. Pond water would have nutrients and other bacteria so that more variables would be added to this investigation other than the two microbes metabolizing the oil.

20. How does the procedure you used in this lab differ from an actual oil-spill cleanup?

Answers will vary but may include the following ideas: Salt water was not used; no existing bacteria or other organisms were in the water; no wave action constantly stirred up the water and oxygenated it; and no sand, rocks, or other shoreline materials were present.

## Conclusions

21. What would happen to the growth of the microbes if no fertilizer were added?

The microbes would still grow and degrade the bacteria but probably at a slower rate. The fertilizer helped coagulate the oil and added nutrients that allowed the microbes to grow and reproduce faster.

**22.** Which microbe degraded the oil better?

Answers will vary. Some students will note that the *Penicillium* culture created more turbid water, but this is due to the formation of hyphae, not increased rate of degradation. In general, both organisms degrade oil at the same rate.

**23.** What advantages, if any, can you think of for using a mixture of microorganisms rather than just one kind of microorganism to degrade the oil?

By using a mixture of microorganisms, a broad spectrum of hydrocarbon degradability can be attained, something that is not possible when using only one type of microorganism.

**24.** An environmentalist may argue that as damaging as an oil spill can be to the environment, adding fertilizer and microbes is worse. Explain this argument.

An environmentalist may argue against adding an organism that is not indigenous to an area. The fertilizer could also encourage growth of other organisms. It may disrupt the balance of the ecosystem.

**25.** If you were an oil-company executive and had to decide how to clean up an oil spill from a tanker, what would you recommend? Explain your methods.

Answers will vary, but students will probably recommend using both mechanical and microbial methods. This would involve first using mechanical skimmers and absorption to immediately pick up as much oil as possible, and then fertilizing the remaining oil and adding microbes. If the spill is around the shoreline, microbes should be used immediately. Students may also suggest ways to avoid putting oil into the water to begin with.

## Extensions

**26.** *Marine biologists* study the relationships of living things and their environment in Earth's seas and oceans. Some marine biologists study the populations and communities that make up a specific ecosystem. Others study the effects of pollution and climate changes on the environment as a whole. Find out about the training and skills required to become a marine biologist.

**27.** Do library research or search the Internet to investigate how effective bioremediation has been in the cleanup of oil spills.

**28.** Conduct this experiment again without adding any fertilizer to the jars. Compare the results of this experiment with the results of the experiment in which fertilizer was added along with the oil.

Name ______________________________

Date ______________ Class ______________

HOLT BIOSOURCES
LAB PROGRAM
BIOTECHNOLOGY

# D12 Experimental Design: Can Oil-Degrading Microbes Save the Bay?

**Prerequisites**
- **Biotechnology D11—Laboratory Techniques: Oil-Degrading Microbes** on pages 57–62

**Review**
- aseptic technique
- bioremediation

PARKS AND RECREATION DEPARTMENT
Eureka, California

February 28, 1998

Rosalinda Gonzales
Environmental Studies Division
BioLogical Resources, Inc.
101 Jonas Salk Dr.
Oakwood, MO 65432-1101

Dear Ms. Gonzales,

I am the Director of Parks and Recreation for Eureka, a town near the Pacific Ocean. Tourism has a significant impact on the local economy. Recently, an oil tanker spilled a large amount of oil about 10 miles from our shoreline. The oil company has done everything it can to contain and clean up the spill, but some of the oil has drifted into our bay.

Last week I was dismayed to see a great deal of oil on the surface of the shallow waters and the beach sand. In addition to being concerned for the local wildlife, I am afraid that the tourist season will begin before we are able to clean up Eureka Bay. Unfortunately, because we are at the end of a slow season, we have a very limited budget for tackling this problem. We need a quick, inexpensive, and effective way to clean up the oil without damaging the local ecosystem.

I recently spoke to a previous client of yours, who recommended your company very highly. I would like your company to help us find a way to clean up the oil without disturbing the ecosystem. I am sending samples of the contaminated water. Please keep me informed of your progress.

Sincerely,

Rebecca Childs

Rebecca Childs
Director of Parks and Recreation

BioLogical Resources, Inc. Oakwood, MO 65432-1101

# MEMORANDUM

To: Team Leader, Ecology Dept.

From: Rosalinda Gonzales, Director of Environmental Studies

I want your department to find the most effective, environmentally sound, and inexpensive treatment for cleaning up Eureka Bay. There are a variety of ways to clean up an oil spill. For example, barriers are often used to contain a spill while skimmers are used to skim oil from the water's surface. Unfortunately, these methods are expensive and can be very time-consuming. I recently spoke with Ms. Childs, and we decided that it would be best to investigate a biological approach to this problem. One such method utilizes bacteria or fungi to break down the oil. Sometimes a cleanup crew will add microorganisms or fertilizer to enhance the growth of microorganisms already present in the soil and water.

Ms. Childs has sent samples of contaminated water and sand from the bay. Have your research teams investigate this approach by comparing the effects of adding the following treatments to the contaminated samples: fungi, bacteria, fungi with fertilizer, bacteria with fertilizer, and fertilizer. Of course, you will also need to perform a control test.

Include in your report to Ms. Childs an explanation of what to expect as the procedure is implemented and what happens to the oil in the process. Also, please add information about the food chain in a typical shallow marine environment so that Ms. Childs can see how your proposed cleanup plan will affect the local ecosystem. Finally, please keep in mind that Ms. Childs is under a very tight schedule and needs these results as soon as possible.

## Proposal Checklist

Before you start your work, you must submit a proposal for my approval. **Your proposal must include the following:**

_____ • the **question** you seek to answer

_____ • the **procedure** you will use

_____ • a detailed **data table** for recording observations

_____ • a complete, itemized list of proposed **materials** and **costs** (including use of facilities, labor, and amounts needed)

**Proposal Approval:** ______________________________

(Supervisor's signature)

## Report Procedures

When you finish your analysis, prepare a report in the form of a business letter to Ms. Childs. **Your report must include the following:**

______ • a paragraph describing the **procedure** you followed to compare the effectiveness of different combinations of microorganisms and fertilizer for degrading refined oil

______ • a complete **data table** including the microbe density and appearance of each sample for each day of growth

______ • your **conclusions** about the comparative effectiveness of each treatment

______ • a detailed **invoice** showing all materials, labor, and the total amount due

## Safety Precautions

- Wear safety goggles, disposable gloves, and a lab apron.
- Glassware is fragile. Notify your teacher promptly of any broken glass or cuts. Do not clean up broken glass or spills unless your teacher tells you to do so.
- Wash your hands before leaving the laboratory.

## Disposal Methods

- Dispose of all waste materials according to instructions from your teacher.
- Place all paper towels and other disposable materials in a trash can.
- Place broken glass and contaminated materials in the separate containers provided.
- Wash reusable materials such as glassware and lab utensils, and return them to the supply area.

## FILE: City of Eureka Parks and Recreation Department

**MATERIALS AND COSTS** (Select only what you will need. No refunds.)

**I. Facilities and Equipment Use**

| Item | Rate | Number | Total |
|---|---|---|---|
| facilities | $480.00/day | ______ | ______ |
| personal protective equipment | $10.00/day | ______ | ______ |
| compound microscope | $30.00/day | ______ | ______ |
| microscope slide with coverslip | $2.00/day | ______ | ______ |
| petri dish | $2.00/day | ______ | ______ |
| test tube with cap | $3.00/day | ______ | ______ |
| test tube rack | $5.00/day | ______ | ______ |
| incubator | $15.00/day | ______ | ______ |
| balance | $10.00/day | ______ | ______ |
| 50 mL graduated cylinder | $5.00/day | ______ | ______ |
| sterile 1 mL pipet with pump | $2.00/day | ______ | ______ |

**II. Labor and Consumables**

| Item | Rate | Number | Total |
|---|---|---|---|
| labor | $40.00/hour | ______ | ______ |
| *Pseudomonas* culture | $20.00 each | ______ | ______ |
| *Penicillium* culture | $20.00 each | ______ | ______ |
| density (turbidity) indicator strip | $2.00 each | ______ | ______ |
| fertilizer | $0.10/g | ______ | ______ |
| distilled water | $0.10/mL | ______ | ______ |
| tap water | no charge | ______ | ______ |
| sterile stirring stick | $1.00 each | ______ | ______ |
| wax pencil | $2.00 each | ______ | ______ |
| water samples | provided | ______ | ______ |
| **Fines** | | | |
| OSHA safety violation | $2,000.00/incident | ______ | ______ |
| | | **Subtotal** | ______ |
| | | **Profit Margin** | ______ |
| | | **Total Amount Due** | ______ |